MW00572176

Advance Praise

'Humanity needs clean energy, and the more, the better – it's the only way to stave off disorder and give us what we need and want. Nuclear power is the only source we know of that is up to this demand, yet irrational dreads and taboos are keeping the world from this potential cornucopia. Marco Visscher engagingly and effectively shows us how we should think about the greatest challenge of this century.'

– Steven Pinker, Johnstone Professor of Psychology at Harvard University and bestselling author of *Enlightenment Now*

'Visscher is right: modern nuclear power will brighten the future for our grandchildren.'

– James Hansen, climate scientist

'It reads like a thriller but informs like a textbook. Essential reading for anyone with questions about nuclear.'

– Mark Lynas, author of numerous books on the environment including *Our Final Warning*

'A lively and even-handed book about a very important subject – the history and future of nuclear power. Original and gripping.'

– Lucy Jane Santos, expert on the cultural history of radioactivity, and author of *Chain Reactions* and *Half Lives*

'Marco Visscher surprises both advocates and opponents of nuclear energy by weaving technology, society, politics and culture into a compelling grand narrative. Whether you like nuclear energy or not you will find plenty in this extraordinary book that will surprise you, and make you think.'

– Rauli Partanen, energy analyst and award-winning co-author of *The Age of Energy*

'A timely intervention into the debate over how to free countries of their dependence on fossil fuels, written by one of the foremost experts on nuclear power.'

– Alexander Kaufman, senior reporter at *HuffPost* with a focus on climate, energy and environment

'Lively and informative, Marco Visscher combines an eye for good stories with a broad expertise on this widely misunderstood energy source.'

– Joshua Goldstein, Professor Emeritus of International Relations at American University, Washington, DC, co-author of *A Bright Future* and co-writer of Oliver Stone's movie *Nuclear Now*

'Decades of scientific studies have shown that the risks from nuclear power are smaller than the risks from energy production by other methods – both for humans and the planet. Yet energy policies appear to be directed by myths, not evidence. This book seeks to address some of these issues and will hopefully lead to better conceived policies before it's too late.'

> – Geraldine Thomas, Professor of Molecular Pathology at
> Imperial College, London, and Director of Chernobyl Tissue Bank

'A clear explanation of why we need far more nuclear power in the coming years.'

> – Oliver Stone, film director of numerous movies including
> *Platoon, Wall Street* and *Nuclear Now*

'This highly accessible book tells the incredible story of radioactivity, atomic bombs and nuclear power. It lucidly explains why so many of us have gotten so terrified, and why this fear is baseless.'

> – Maarten Boudry, philosopher of science, first holder of the
> Etienne Vermeersch Chair at Ghent University, Belgium

'In this timely and important book, Marco Visscher skilfully provides clear-eyed histories of accidents at Chernobyl and Fukushima, debunking outrageous claims about the dangers of nuclear power while describing the nuclear resurgence. Highly recommended.'

> – Robert Bryce, author of *A Question of Power*, co-producer
> of the docuseries *Juice: How Electricity Explains the World*

'This masterly and entertaining account of the past and present of nuclear energy shows what we can learn from Fukushima and how we should react to nuclear accidents (if any were to occur in the future) which is particularly important as nuclear energy is likely to become an even more important part of the global energy mix.'

> – Simon Friederich, associate professor of philosophy of
> science at University of Groningen, Netherlands

'In this era of dangerous climate change, we are all paying a heavy price for having let irrational anxiety over nuclear energy turn us away from one of the world's most powerful decarbonising tools. In this stimulating book, Marco Visscher unpicks in brisk prose exactly how we reached this impasse, and how we might respond.'

> – Jonathan Symons, Senior Lecturer in Politics and International
> Relations at Macquarie School of Social Sciences, Sydney, and
> author of *Ecomodernism: Technology, Politics and Climate Crisis*

'Using a refreshingly optimistic and engaging voice, ...Marco Visscher offers a great analysis of the many misconceptions that have stalled nuclear energy's growth for decades, ...showing why nuclear is the obvious choice for tackling climate change and providing energy security.'

> – Bret Kugelmass, CEO of Last Energy and podcast host at *Titans of Nuclear*

THE POWER OF NUCLEAR

Also Available

THE POWER
OF NUCLEAR

THE RISE, FALL AND RETURN
OF OUR MIGHTIEST
ENERGY SOURCE

Marco Visscher

BLOOMSBURY SIGMA

LONDON · OXFORD · NEW YORK · NEW DELHI · SYDNEY

BLOOMSBURY SIGMA
Bloomsbury Publishing Plc
50 Bedford Square, London, WC1B 3DP, UK
29 Earlsfort Terrace, Dublin 2, Ireland

BLOOMSBURY, BLOOMSBURY SIGMA and the Bloomsbury Sigma logo are
trademarks of Bloomsbury Publishing Plc

Originally published in Dutch as *Waarom we niet bang hoeven te zijn voor
kernenergie: De emoties & de feiten* by Nieuw Amsterdam, 2022

First published in English by Bloomsbury Publishing Plc, 2024
as *The Power of Nuclear*
English translation © Marco Visscher, 2024

Copyright © Marco Visscher, 2024

A catalogue record for this book is available from the British Library

Library of Congress Cataloguing-in-Publication data has been applied for

ISBN: HB: 978-1-3994-1907-9; eBook: 978-1-3994-1906-2

2 4 6 8 10 9 7 5 3 1

Typeset in Bembo Std by Deanta Global Publishing Services, Chennai, India
Printed and bound in Great Britain by CPI Group (UK) Ltd, Croydon CR0 4YY

MIX
Paper | Supporting
responsible forestry
FSC® C171272

To find out more about our authors and books visit www.bloomsbury.com and
sign up for our newsletters

To all the wizards

'In some sort of crude sense which no vulgarity, no humour, no over-statement can quite extinguish, the physicists have known sin; and this is a knowledge which they cannot lose.'

– J. Robert Oppenheimer, Arthur D. Little
Memorial Lecture, 25 November 1947

Contents

Prologue

Rebellion

'In our reasonings concerning matter of fact, there are all imaginable degrees of assurance, from the highest certainty to the lowest species of moral evidence. A wise man, therefore, proportions his belief to the evidence.'

– David Hume, philosopher, in
An Enquiry Concerning Human Understanding, 1748

He's not exactly in a good mood. It's odd, really. Here is one of the first and most authoritative scientists to warn about climate change, finding himself at the 2017 UN Climate Change Conference in Bonn, Germany, where everyone is coming together to agree on lowering carbon emissions. This must be heaven!

Yet James Hansen is anything but thrilled. 'It's just words.'[1]

His signature hat is on the table between us. 'Politicians say we have to prevent catastrophic climate change,' Hansen continues. 'But a conference like this makes no difference. Government leaders pat each other on the back and smile politely for the camera. But all those words? Bullshit.'

It's safe to say the world's leading climate scientist is a little impatient.

He has every reason to be. Since the Kyoto Protocol in 1997, every climate treaty has had a negligible effect on greenhouse gas emissions. Despite all the promises, the concentration of carbon in the atmosphere has risen sharply. It seems we will have to learn to live with the consequences. According to Hansen, these consequences won't be mild.

The choice of Germany as host for the UN conference should be promising. Here, wind turbines and solar panels are making unprecedented advances. Already, the German government has spent hundreds of billions of euros on these renewables.[2] The host country is considered a model nation.

Hansen does not agree. 'Obviously, power from solar and wind is useful,' he explains, 'but cannot yet be stored long-term or affordably for when the weather doesn't cooperate.' In such cases, it's usually power plants using natural gas or coal that need to step in. Thus, moving away from fossil fuels by mainly using weather-dependent power is, in Hansen's opinion, 'not a good strategy'.

The figures prove him right. In recent years, carbon emissions from the German electricity supply have barely declined[3] and the country has some of Europe's most polluting coal-fired power plants.[4] Indeed, in the year before the climate summit in Bonn, carbon emissions went up.[5] Germany's carbon emissions per capita are above the European average.[6]

And just who is James Hansen? The son of a farmer, he was born in 1941 in a small town in Iowa, in the corn-producing heart of the United States. While studying astronomy and physics, Jim developed an interest in the dust clouds surrounding Venus and landed a job at NASA. There, at the space agency, his attention shifted.

The ozone layer in the atmosphere was being affected by chemicals used in everything from refrigerators and air conditioning to foam plastic and aerosol spray dispensers. All of this contributed to a greenhouse effect – an effect that, as Hansen learned, came mainly from burning fossil fuels. What impact would that have on his home planet?

He started tinkering with a program on what was then the world's largest computer. During long days in the NASA lab in the heart of New York – just a few floors above Tom's Restaurant, the eatery made famous by the sitcom *Seinfeld* – he developed one of the first climate models. The scientist became alarmed.

Hansen started publishing papers and giving presentations. But it was only when he was invited to address the US Congress in 1988 that climate change entered the public's mind. His message: the Earth is getting warmer. Global warming is already taking place.

At that time, Hansen was head of the Goddard Institute for Space Studies, the NASA department dedicated to atmospheric change. He would remain so until he stepped down in 2013. By then, he had a hefty stack of studies to his name, which were held in high regard by peers.

As Hansen watched the evidence for man-made global warming grow, he became frustrated with climate policy. Fossil fuels had been exposed as the biggest culprit. Yet coal, oil and natural gas provide up to 80 per cent of global energy consumption – a share that has barely declined in the last 40 years.[7] Electricity from solar and wind may be on the rise, but in the 2022 global energy mix, their combined output sat under 5 per cent.[8]

When politicians kept muddling along and he himself became a grandfather, Hansen realised: *We need to do more, and fast.*

Thus, the scientist became an activist. Hansen found himself handcuffed at protests against fossil fuels. He proclaimed that top executives at oil companies should be tried for high crimes against humanity and nature.[9] He compared coal transports to 'death trains' heading to concentration camps.[10]

And now, at the climate summit in Bonn, Hansen sees the future as bleak. 'We are facing an emergency,' he says. 'If we don't come up with a source of reliable zero-carbon energy soon, our children and grandchildren will have to. They will have even less time to repair the damage we cause.'

Thankfully, solutions exist. Just before our conversation, at a press conference, Hansen talked about one of those solutions – nuclear power.[11] From an objective point of view, this makes perfect sense. A nuclear plant doesn't emit any greenhouse gases, and provides electricity 24/7.

In addition, as Hansen learned, there aren't that many success stories about weaning off fossil fuels and lowering carbon emissions. 'But the times when countries were able to produce a lot of new zero-carbon energy in a short time,' he says in Bonn, 'they did it with nuclear power.'

Apparently, that is an inconvenient truth to some. At the press conference, anti-nuclear activists, gathered together in one of the front rows, shook their heads ostentatiously. Once Hansen's presentation was over, they grabbed the microphone and started asking questions. Didn't Mr Hansen know nuclear power had become incredibly expensive? Where did he want to store the waste for the next tens of thousands of years? Surely, in a modern democracy, nobody in their right mind would long for a revival of nuclear power?

Hansen is used to such resistance. Throughout society, nuclear power is met with deep-rooted suspicion. For many, nuclear has something evil about it, something shady. To these people, there is a mysterious, ominous feel to it, a threat of imminent danger. It's almost as if the fission of atoms, as in a nuclear reactor, does not belong in this world.

Objections to nuclear power are well known. An accident could make large areas uninhabitable. The radiation released might cause diseases and deformities far and wide. At present, there's no way to safely store the waste for tens of thousands of years. The construction of a nuclear plant – averaging six and a half years between 2000 and 2021,[12] but running at well over a decade for new and upcoming reactors in the United Kingdom, France and Finland – simply takes too long to mean anything in terms of reaching climate goals. A terrorist gaining access to a nuclear plant could make nuclear weapons, or blow it up…

All these objections are easy to disprove, Hansen knows. But he also realises that not everyone is easily convinced. Now that he runs a small organisation that depends on donations, Hansen notices some of that rigid resistance. The vast majority of potential benefactors support his protests against oil pipeline construction and his calls for an international carbon tax, but when they hear that James Hansen, a hero to them and so many others, thinks nuclear power is actually a good idea, they flinch. *Nuclear power?!* It often leads to them not wanting to support his work.

Hansen shrugs. 'So be it.'

Like no other, Hansen knows the conversation about nuclear power is challenging. It's also a conversation that's inevitable. The role of nuclear power is not at all over. Around the world, some 440 nuclear reactors are in operation, spread across more than 30 countries.[13] In Europe, no other source produces more electricity, with one in four light bulbs burning thanks to a nuclear plant.[14]

It doesn't stop there. As of 2024, over 60 reactors are under construction, over 90 reactors are planned, and over 340 are proposed.[15] China alone announced in 2021 that it wants to build as many as 150 nuclear reactors in 15 years.

The interest in nuclear power is not only because the climate is changing; the world is changing as well. Even before Vladimir Putin sent his army to invade Ukraine in February 2022, commentators pointed to the dangers of strong dependence on fossil fuels from Russia. With nuclear plants, which run on uranium that can be found on every continent, society has a continuous source of zero-carbon energy.

Moreover, the demand for energy will increase significantly in the coming decades. As people in poor and emerging countries aspire to a better life, they will use a lot more energy to meet their needs. For their own well-being, it is better if that energy comes from power plants that do not pollute the air or disrupt the climate.

Yet it is too early to say that nuclear power is on the rise. Between 2000 and 2023, a total of 117 nuclear reactors were connected to the grid. However, in the same years, 121 have been permanently shut down.[16] In the past 25 years, nuclear power's share of the global electricity mix has plummeted from 17 per cent in the 1990s to less than 10 per cent.[17] A number of countries are determined to abandon it for good. Elsewhere, political support is fragile. An accident – not unthinkable, despite assurances of pro-nuclear advocates – could end construction plans in the blink of an eye.

The Intergovernmental Panel on Climate Change (IPCC), the United Nations climate science body, indicates that the pace of nuclear power expansion is being 'constrained by social acceptability in many countries due to concerns over risks of accidents and radioactive waste management'.[18]

James Hansen likes to make a comparison with Galileo Galilei, the Italian astronomer who realised 400 years ago that the Earth moves around the sun. The authorities told Galileo that he had better agree with their view of the Earth as the centre of the universe. Galileo backed down. His silence made life a lot easier for him and he knew that one day his findings would surface anyway.

'But today,' says Hansen in Bonn, 'we cannot remain silent. If we sit back and say that in a few decades' time it will become clear that phasing out fossil fuels will not succeed without nuclear power, we will be right, but by then, it will be too late!'

James Hansen – the scientist, the activist – cannot help but tell it like it is. That's what he does when talking about the climate, and that's what he does when talking about nuclear power. 'The opposition to nuclear power is truly insane,' he sighs. 'All these fears – about radiation, about waste, about accidents – have no basis in science. This aversion is quasi-religious and irrational.'

Is it, really?

PART ONE
BOOM!

Pandemonium

Why did we have to split atoms?

'The question can be raised whether mankind benefits from knowing the secrets of Nature, whether it is ready to profit from it or whether this knowledge will not be harmful for it.'

— Pierre Curie, physicist, speech on acceptance of the Nobel Prize in Physics, 6 June 1905

At a small-town fair in Florida, 12-year-old Paul Tibbets Jr was allowed to go for a ride in a stunt plane. Flying over the sandy shores of the new resort area of Miami Beach, he threw out packets of candy bars that swirled down on a paper parachute. It was a beautiful sight. What an adventure to be so high above the ground! Suddenly the little boy knew: he wanted to be a pilot.

Paul went on to become a pilot, and not just any pilot. He became the finest in the US Air Force. In World War II, he flew the lead bomber when the Americans first appeared over Europe in broad daylight, and led the invasion of North Africa, in which the Nazi-affiliated French Army were chased out. Then, in December 1944, Paul Tibbets, now 29, was put in charge of the unit tasked with deploying a secret weapon – one that could end the war.

On 6 August 1945, Tibbets was at the wheel of a B-29 for a historic mission over Japan. He knew exactly what to do. Endlessly he had practised the sharp turn he had to deploy as soon as the cargo was unloaded. If he executed that manoeuvre perfectly and quickly gained speed, he would, presumably, be safely away when the explosion came. His life depended on it.

Two days earlier Colonel Tibbets and his crew received a briefing. They learned of their special cargo: the most destructive bomb ever created. The men, almost all in their twenties, did not know what to say. They stared at their shoelaces uncomfortably.

The weaponeer gave no details about the nature of the weapon, but told them about the test of a similar model. In the US desert, that single bomb had made a crater more than 300m (1,000ft) deep. Two hundred and fifty kilometres (150 miles) away, windows shook in their frames. Closer, but still 30km (20 miles) away, a number of experts watched. One, not wearing sunglasses, was blinded for half a minute by the bright flash in the dead of night. Another was surprised by the shock wave and was blown over, 40 seconds after the blast. The dust cloud travelled 12km (7 miles) into the air.

The weaponeer's words impressed Tibbets' men. They couldn't know that right after the explosion, the man who later gained fame as the 'father' of this bomb, J. Robert Oppenheimer, was reminded of words from the ancient epic *Bhagavad Gita*: 'Now I Am Become Death, Destroyer of Worlds'.[1]

En route to Japan, with the Pacific Ocean below him, Tibbets was tired. Coffee and ham sandwiches helped him stay awake. He and his crew had risen early. Before take-off from the air base on the small island of Tinian at 2.45 a.m., after a breakfast of eggs and pineapple fritters, they had pictures taken in front of the Boeing. As one would later recount, it was like being at a Hollywood premiere.

The co-pilot took over so that Tibbets could rest his eyes, but he was restless and crawled to the back to chat with the others. Did they have any idea what kind of bomb it was? There was some guessing, joking and chuckling. Then it dawned on someone: 'Colonel, are we splitting atoms today?'[2]

Bingo.

A little after 8 a.m., with Tibbets back in his seat, they were flying over Hiroshima, a city of over 300,000 people. It was a sunny day. Then the target, the Aioi Bridge in the city centre, came into view. And there it went: the first atomic bomb dropped on enemy territory. It fell from almost 10km (6 miles) above ground. The plane, suddenly over 4,000kg (4 tons) lighter, swung upwards. Tibbets jerked the steering wheel to the right for the difficult turn he had practised so much and gave it full throttle. *Out of here!*

The bomb, now tumbling downwards, was equipped with an ingenious radar system that activated the detonation at 580m (2,000ft) above ground for optimal destruction. Forty-four seconds after release, that moment arrived. An explosive caused two lumps containing a combined 64kg (140lb) of enriched uranium to collide. Now the miracle happened: atoms split.

And then? Just 0.7g (0.025oz) of the uranium – no more than the weight of a butterfly – was converted into energy. A fireball was created, with a diameter of up to 300m (1,000ft). In one fell swoop, countless people were scorched. They evaporated, leaving a shadow in the concrete, or were turned into a pile of smoking ash, or left in office buildings like skeletons in their office chairs, sometimes with only a watch around their wrist bones.

Then came the shock wave. Buildings collapsed, kerb stones flew around. People were buried under the rubble, or hurled away, straight through windows, their bodies torn to shreds.

A firestorm charred everyone nearby. Faces, coloured black or red, swelled. Their hair was gone. You couldn't see where their eyes and mouth were. Front and back were indistinguishable.

Some screamed, others moaned. Children cried for their parents, parents cried for their children. The survivors were naked or had only shreds of clothing around them. Skin came off and hung on their bodies like a rag.

Soon, black rain full of radioactive soot descended.

High in the sky, the B-29 was on its way back home. A shock wave gave the tail a solid thump. The plane creaked. In the front, Tibbets noted with satisfaction that he had been fast enough to get away. Behind him, at the opposite end of the plane, the tail gunner couldn't believe what he was seeing. A cloud full of pulverised debris erected itself menacingly in the shape of a mushroom. Only 600km (370 miles) away, the cloud was no longer visible.

The co-pilot noted in his logbook: 'My God, what have we done?'[3]

The atomic bomb did not just fall from the sky; for some time, it was in the horror chamber of our imagination. But where did it come from? The idea for such a weapon first arose after bright minds proclaimed that an awe-inspiring amount of energy was locked up inside the atom.

In around 1900, the sleepy world of physics was being shaken up. Something seemed to be up with the atom. For at least 2,000 years, the atom had been held to be the smallest building block of all matter, with the Greek *a-tomos* (from which the word is derived) meaning 'indivisible'. An atom is about a million times smaller than the diameter of a hair on your head. Every atom, it was thought, has a certain size, shape and mass, and that never changes. Every new substance is just a rearrangement of atoms.

Now, that image wavered. There was talk of mysterious beams called X-rays, which could somehow penetrate through solid materials. It seemed that uranium naturally emits these invisible yet powerful rays. But how? Was the radiation the work of atoms, doing something previously thought impossible?

There was plenty to discover. Young scientists smelled opportunity and uncovered a miniature world. Every atom, it turned out, has a core, called a 'nucleus', surrounded by flying electrons. These have a negative electric charge and are attracted to the positively charged protons in the nucleus. The structure is minuscule. Suppose each of the atoms in a grapefruit is the size of a blueberry. In that case, the grapefruit would need to be the size of the Earth to hold them all. If you want to know how small the nucleus of an atom is, blow up one of the blueberries to the size of a big football stadium, get a pair of binoculars and look for something the size of a marble.[4]

One of the scientists behind these discoveries was Ernest Rutherford. As a boy, he had helped with chores on his parents' farm in New Zealand: he planted potatoes, chased birds away from the fruit trees and drove the cows home when it was time for milking. He won a scholarship, excelled at school and was admitted to the University of Cambridge in the UK in 1895. Rutherford would return there, with a Nobel Prize in his pocket, as professor and director of the Department of Physics in 1919. By the time he died in 1937, this farmer's son was widely recognised as the pioneer of the atomic age.

Rutherford earned that recognition when he discovered in 1900 that thorium, an element that emits radiation itself, gives off a kind of exhalation, a gas. He needed someone with an understanding of chemistry and teamed up with the younger Frederick Soddy, who determined that the exhalation is helium, a colourless noble gas. Together, they figured out what was going on here: the transmutation of atoms.

Transmutation? That sounds like the dream of ancient alchemists. During the Middle Ages, alchemists used liquids, powders, potions, bottles and furnaces in an attempt to transform base metals into gold, bestowing immortality on humanity.

Church leaders warned against these sorcerers who tried to unravel the sacred mysteries of the universe. Such revolutionary ideas had to be outlawed. In the end, it was modern chemistry that dismissed alchemy as an occult activity for charlatans. Transmutation would always remain an illusion.

Well, not quite, or so it seemed now. Certainly, the cautious Rutherford dared not use the word. 'Don't call it *transmutation*,' he told Soddy. 'They'll have our heads off as alchemists.'[5]

But perhaps the ancient sorcerers were on to something. Rutherford and Soddy's discovery showed that nature itself can cause an element to change into another over time. In that process of decay, they theorised, energy is given off – an energy hidden in a tiny atom.

News of the atom surfaced at a tumultuous time, when energy was at the root of all sorts of changes. Cars displaced horses and carriages. Steamships sailed the oceans. There was talk of flying in an aircraft. The big city with its electrical lights beckoned – this was where you could find a job and go to the movies. The advance of technology and science made the world more cheerful. With all that energy inside the atom, a clean and prosperous world seemed within reach.

A new discipline – nuclear physics – began to form, which would be showered with Nobel prizes for years to come. And yet despite these awards, the winners had to acknowledge their ignorance: experimental physicists could not always explain their findings, and the theorists among them struggled to demonstrate their ideas. No one yet knew *if* or *how* to liberate atomic energy.

But something else was considered, too: if the energy could be liberated, it could be used for explosions – big explosions. In 1903, Rutherford wrote in his notes: 'Some fool in a laboratory might blow up the universe unawares.'[6] In a lecture to a British Army corps in 1904, Soddy reported that whoever could break open the atom and harness the released energy would have 'a weapon by which he could destroy the earth if he chose'.[7]

Nuclear physics seemed both exciting and scary. Some questioned whether, as the atom is so deeply hidden from us, it shouldn't stay that way, concerned we might be robbing nature of its secrets.

Then war broke out. The conflict began in Europe in 1914 and spread to Africa, the Middle East and Asia. Seventy million soldiers enlisted and were drafted. Ernest Rutherford and Frederick Soddy saw students and staff leaving for the front line. Not everyone returned.

Old-fashioned warmongering mixed with modern-style technology. For military consultations, the carrier pigeon was replaced by field telephones. Weapons were becoming more efficient: machine guns, mortar shells, fragmentation bombs. Chemical weapons like tear gas, chlorine gas and mustard gas became symbols of the new ingenuity. The war lasted four long years and claimed 20 million lives.

Once the smoke of the Great War had cleared, a blanket of pessimism fell over Europe. It seemed that science and technology did not solve everything after all, yet together they continued to boom. Rutherford theorised on the existence of neutral, electrically uncharged particles in the atomic nucleus that seemed quite useless. A colleague exposed the atomic nucleus of nitrogen to radiation and got other particles as a result, becoming the first to deliberately transmute one element into another.[8]

Their work did not go unnoticed. Journalists from popular magazines wrote about the wonders coming from the laboratories. Nuclear physicists were so often called the 'new alchemists' that it became a cliché. Writers discovered an inexhaustible source for their science fiction stories.

Expectations were high. In 1926, Leon Trotsky, the Russian revolutionary, spoke out on the issue. In front of an audience, he asserted: 'The atom contains within itself a mighty hidden energy, and the greatest task of physics consists in pumping out this energy, pulling out the cork so that this hidden energy may burst forth in a fountain. Then the possibility will be opened up of replacing coal and oil by atomic energy, which will also become the basic motive power. This is not at all a hopeless task. And what prospects it opens before us!'[9]

Meanwhile, things kept on brewing in Europe. Traditions were being shaken, old strongholds were crumbling. As empires fell, democracy was thriving. Radical political ideas were established – communism in Russia, fascism in Italy.

During these years, Göttingen University in Germany was the intellectual centre of atomic science. This is where a young physicist named Werner Heisenberg laid the foundations for quantum mechanics, the study of particles smaller than atoms. More articles in professional journals appeared in German than in all other languages combined.[10]

Like a magnet, Göttingen attracted students from all over the world. Among them was the man who would lead the creation of the atomic bomb: Robert Oppenheimer, a tall, slim student suffering from severe depression. This young American read poetry and, for a better understanding of the Hindu epic *Bhagavad Gita*, learned Sanskrit.

Another scientist – the most famous of all – made the reverse crossing. Albert Einstein left Europe for the United States because of the rise of anti-Semitism in Germany. Jews were blamed for everything. Adolf Hitler, an outraged former soldier, dreamt of a superior Aryan race in need of *Lebensraum* (living space). A new world war was not unthinkable.

Ernest Rutherford, too, was worried, though he tried not to show it. When a journalist asked if nuclear fission could be a useful way of obtaining energy, he said 'Never.' The nucleus of the atom would always remain 'a sinkhole', not a source of energy.[11] At a lecture in 1933, Rutherford stated that 'a lot of nonsense' was being said about the possibility of obtaining atomic energy. It would not be possible, he said, to control it in such a way that 'would be of any value commercially, and I believe we are not likely ever to be able to do so'.[12]

The eminent scientist did not seem to realise the atom's true potential. Or perhaps he didn't want to admit it, and face the consequences of harnessing the energy of the atom. Indeed, based on correspondence and the diary entries of his peers, it appears Rutherford sensed that the secret of the atomic nucleus must be unravelled at some point.[13] With the threat of war looming, he was unwilling to think about the consequences, and he certainly did not want to live through them. Saddened by the thought that his beloved atom would inevitably be used for destruction, he decided on a policy of discouragement. Rutherford, by now recognised as a baron in honour of his scientific achievements, would go on to publicly proclaim that it was impossible, even an absurd idea, to ever capture that energy.

However, it was precisely his stubborn denial of any practical application that would lead to the atomic bomb...

Leó Szilárd was an eccentric intellectual: quirky, brilliant, arrogant. He preferred to be lost in thought all day, taking walks or in the bath, brooding on technical innovations, musing on how he would change the world.

After studying physics in Berlin, Szilárd came up with a series of inventions that he went on to patent. He was able to support himself with the proceeds. He became friends with his former professor Albert Einstein and together they filed dozens more patents. Some of these concerned a refrigerator that did not need electricity, only a heat source, to run. Electrolux bought the rights, though it never became a commercial success.

In 1933, after Hitler was appointed Chancellor, Szilárd arrived in London from Germany. It was the second time he had fled anti-Semitism, having already left his native Budapest in 1919 after the Bolsheviks took power in Hungary. So there he was, sitting in the lobby of the Imperial Hotel in London, reading a newspaper. An article about a scientific conference in Leicester on the state of nuclear physics caught his eye. It will not have done his mood any good that he himself, the great Leó Szilárd, had not received an invitation. He started reading.

The venerable Lord Rutherford was mentioned. He spoke of 'a very poor and inefficient way of producing energy'.

It was unlikely, Rutherford said, that it could be done. Anyone who saw in the transformation of atoms a source of energy, he argued, was just 'talking moonshine'.[14]

Szilárd looked up from his newspaper. Talking moonshine? What a ridiculous thing to say... He decided to go for a walk.

That's when it happened. While strolling the streets of London, Szilárd had a brainwave. Neutrons, those useless particles in the atomic nucleus, might have a function after all. One could use them to cut right through an atom. If there were an element, Szilárd pondered, that, when splitting the nucleus of the atom, was left with a neutron that in turn can collide with another atomic nucleus that also splits, resulting in another neutron, then... Yes! The process maintains itself! *Then you'd have nuclear power!*

It was a classic eureka moment. On the morning of 12 September 1933, crossing Southampton Row near the British Museum in Bloomsbury, 35-year-old Leó Szilárd invented the nuclear chain reaction. The energy released would be millions of times greater than in any chemical reaction. It would guarantee an abundant source of energy in just one small package.

His insight would become the basis of the explosive power in the atomic bomb and of electricity from nuclear plants. Szilárd was to change the world with it.

But first, he applied for a patent.

Many questions remained. Such as: which element would yield two neutrons if one is captured? Does such a thing even exist? For an answer, you would have to systematically subject all of the elements to experimentation in a laboratory: an expensive, boring, time-consuming job for which Szilárd had no appetite. He travelled to laboratories to see if there was any interest, but nobody saw much point in such a project. Even Rutherford turned Szilárd away when he learned that his visitor had dared to file a patent.

Soon, however, among scientists, the idea of a nuclear chain reaction came to life. It was being thought about and experimented with. Was it indium? No. Beryllium then? Not that either.

Days, months, years went by. Nothing seemed to work. Szilárd, now living in the United States, lost faith. In late 1938, disillusioned, he sent a letter to the British Admiralty, to whom he had given custody of the patent as a military secret so his idea wouldn't fall into the wrong hands. Szilárd wrote asking them to forget about the whole thing.

Just as that letter was on its way, scientists in Europe reported a breakthrough. Otto Hahn and Lise Meitner claimed success with the very last element in the periodic table: uranium, which, with 92 protons in its atomic nucleus, is the heaviest element found in nature. After they bombarded the atomic nucleus of uranium with neutrons, the remnant contained traces of other elements. Mass was transformed into energy. The miracle of alchemy had been witnessed. They called it 'nuclear fission'.

Who were these scientists? Hahn was a German who, as soon as war broke out, started working on the Nazi programme for an atomic bomb, just as he had helped to develop poisonous gases like chlorine and mustard gas in the previous war. Meitner was an Austrian who would be written out of the history books by Hahn; she came from a Jewish family and later would narrowly escape the Nazis when boarding a train to the Netherlands. Together, they laid the foundations for the atomic bomb.[15]

Immediately after Hahn and Meitner's discovery of nuclear fission, Szilárd wrote to the British Admiralty about his patent, asking them to please ignore his previous letter. Szilárd knew what the discovery meant. After years of theorising and experimenting, it had finally been shown that mass could be converted directly into energy and trigger a nuclear chain reaction. Now, nuclear power was inevitable. And whoever succeeded in harnessing nuclear fission on a large scale was not going to make electricity, but an explosive more powerful than had ever been seen before. There was also agreement that whoever got such a bomb first would win the next war.

This realisation resulted in Szilárd paying a visit to his old friend Einstein at his summer cottage on New York's Long Island in 1939. Einstein had become an icon. Szilárd told him that now was the time for him to leverage his reputation so the Americans would have an atomic bomb before Hitler. Szilárd was sure the Germans were already working on one. After all, they were at the forefront of nuclear fission research.

His host interrupted. An atomic bomb? Einstein, in a bathrobe and slippers, was perplexed. 'I did not even think of that.'[16] All those years, Szilárd had hardly thought of anything else.

When their joint letter reached the White House through a friend of a friend, the German army had overrun Poland with tanks and dive-bombers. The airstrikes on civilian targets in particular were met with horror. US president Franklin D. Roosevelt said that these attacks had 'profoundly shocked the conscience of humanity' and spoke of 'inhuman barbarism'.[17] France and the United Kingdom declared war on Germany. The United States joined the battle two years later, after Japan sought to increase its influence in Asia and launched an air attack on Pearl Harbor, the US naval base in Hawaii.

Yet before they joined the war, the Americans did not stand aside. Roosevelt recognised the threat flagged by Szilárd and Einstein, and ordered the development of an atomic bomb in October 1941. Initial scepticism among the army leadership for such a completely new weapon was slowly overcome. In turn, the scientists slowly grew used to being in the service of the US armed forces.

With unprecedented speed, an entire industry was created. The Americans, never the best at fundamental research and always more focused on practical applications, benefited from the knowledge of dozens of top nuclear physicists who had fled Nazi Europe.

One was Enrico Fermi, an Italian called an 'architect of the atomic age' after his death in 1954.[18] Fermi was a genius. He made complex calculations as fast as lightning and without error. His colleagues in Italy called him 'the Pope' because of, as author James Mahaffey later put it, 'his apparent spiritual connection with the higher deity of physics'.[19]

Fermi's wife was Jewish. After Mussolini introduced anti-Semitic laws in October 1938, following the German example, mixed marriage was banned. On 10 November, the day after *Kristallnacht* – a pogrom during which Nazis attacked Jews across Germany and destroyed their homes, shops and synagogues – the phone rang: Fermi was to be awarded the Nobel Prize in Physics. A month later, he and his wife and children collected the award in Sweden before travelling on to America with the prize money. And so another outstanding European scientist was welcomed to the New World.

Fermi, who had studied in Göttingen, was put in charge of an experiment to prove a nuclear chain reaction. On the wooden floor of a squash court under the stands of an abandoned American football stadium at the University of Chicago, he and his team started building a nuclear reactor. To improve his English, the Italian read *Winnie the Pooh* in his spare time, and gave his tools and instruments names like Piglet, Tigger and Heffalump.

The oval construction of the reactor filled pretty much the entire squash court. Blocks of graphite were used to slow down the newly created neutrons so they were better able to split the nucleus of another uranium atom. To manage the chain reaction, Fermi had control rods installed: plates of cadmium, an element that can quickly capture neutrons. At each subsequent step, these control rods would be removed, one by one, until a self-sustaining nuclear reaction occurred. Should things go haywire, the control rods could be quickly pushed back.

The demonstration took place on 2 December 1942. Chicago was experiencing its coldest day in half a century.

From the stands, just under 50 spectators watched breathlessly, dressed in winter coats, wearing hats and gloves, freezing in the unheated venue. The experiment succeeded. The nuclear chain reaction, secretly generated without any protective measures in the centre of America's second city, occurred exactly as Fermi had calculated. The ticking of radiation detectors made onlookers nervous. When, after a few minutes, an icy calm Fermi decided that all was working well, the control rods were pushed in and the whole thing was shut down.

Afterwards, the scientists shared a bottle of Chianti, drinking it from paper cups. One of them was Leó Szilárd. He had collaborated with Fermi and together they were to draft the patent application for the nuclear reactor. But Szilárd struggled with mixed feelings. It worked, he knew, but he realised something else. When the two of them were left alone, he shook Fermi's hand and said: 'I think this day will go down as a black day in the history of mankind.'[20]

Now things started moving fast. Of the mined uranium, less than a measly 1 per cent appeared to be of use. Only that part – the isotope called U-235 – could act as fissile fuel and cause a chain reaction. Many tonnes of seemingly worthless rock were carved out, processed and purified to eventually extract a few pounds of fissile metal. Nothing was wasted. Experts claimed there was not that much uranium on Earth.

Everything was done in secret. Entire cities sprang up, hastily built in sparsely populated landscapes. They didn't appear on maps and nowhere along the road did signs point the way. The tens of thousands of laboratory and factory workers were kept in the dark about what exactly they were working on. Everything was surrounded by barbed wire, watchtowers and guarded gates. There were no more than a handful of telephone lines, all of which were monitored. All incoming and outgoing mail was checked.

Everyone was suspect, even Oppenheimer, who was appointed scientific director of what came to be known as the

Manhattan Project. He had no managerial experience and did not have a Nobel Prize. His wife, his mistress, his brother and his sister-in-law were all members of the Communist Party. He himself held rather left-wing views. The FBI followed him closely, checking out all the scientists Oppenheimer persuaded to participate. Appealing to the scientists' patriotism, he convinced them to leave the safe environment of their university or laboratory. Working on something important and secret in wartime proved irresistible.

To make that tiny percentage of fissile material from uranium work in a bomb, it had to be enriched, a process whereby the percentage of U-235 is gradually increased by concentrating it. It was a laborious job, which took place using powerful machines in huge factories. Day and night, thousands of women from surrounding villages commuted back and forth in buses. During their shifts, each woman sat on a stool and peered at a needle that had to point straight up on a gauge. At the slightest deviation, she had to turn a knob to correct it. That was it. Posters reminded them not to discuss their work with anyone. That cannot have been very difficult; they had no idea they were enriching uranium.

Another route to make an atomic bomb is through the production of plutonium, which is also fissile. This element does not occur in nature. It was discovered in a laboratory in 1940 as a by-product created when uranium is bombarded with neutrons. The Americans made it by irradiating uranium in nuclear reactors for several weeks, causing some of it to transmute into plutonium. A lump of plutonium as small as a tennis ball can be compressed by a shell of common explosives for a blast that is even more powerful than one made with enriched uranium.

To play it safe both routes to the atomic bomb were taken.

On the other side of the ocean, World War II was in full swing. Once again, warfare was becoming more efficient. Bombs flattened cities. From the Dutch coast, thousands of

V2 rockets were fired at London for months by German troops. Jews were deported to concentration camps and gassed with Zyklon B, an insecticide developed by the chemist Fritz Haber. A ruthless bureaucracy supported the armed forces. In less than six years, the war claimed some 75 million lives, most of them civilians.

The Americans, together with the British and Canadians, were not the only ones working on an atomic bomb. The Germans were developing one too, much to the Americans' concern. The Japanese were also joining in, led by a former student of Ernest Rutherford. The Russians were kept up to date by spies in the United States.

Only a handful of people in the US military knew exactly what was going on. The confidentiality surrounding the Manhattan Project was taken extremely seriously. This became evident when Roosevelt suffered a stroke and died in his retreat on 12 April 1945. His replacement, Vice President Harry Truman, only found out about the atomic bomb in his initial briefing.

That same evening, as the American people mourned the sudden death of their popular president, final measurements were taken on what was coded Little Boy. The atomic bomb was ready.

Was the world also ready?

The question of whether the bomb should still be used emerged. There was evidence that the Germans hadn't got very far with their *Atombombe*. In Europe, the war seemed almost over. Before the end of the month, Mussolini was captured and killed during an escape attempt, and Hitler had shot himself through the head in his bunker in Berlin. On 8 May 1945, Nazi Germany surrendered to the Allies.

Only Japan remained. It would not be easy to beat them. Notoriously fanatical, the Japanese were known for their brutal war crimes as much as their refusal to surrender. When the Japanese ran out of ammunition, kamikaze pilots flew their planes straight into American ships. Civilians dug tunnels and would rather die than surrender. Children were taught how to strap explosives to their chests during an invasion, lie flat and ignite the thing once a US tank drove over them. This could have been a long, bitter battle. Perhaps they hoped the Americans would give up the fight, having lost more than 100,000 soldiers in Europe.

Then pilot Paul Tibbets began his historic flight...

Three days later, Hiroshima was still burning, the death toll from that one bomb running at some 70,000. Japanese army leaders were just beginning a consultation when news reached them of yet another bomb. This time Nagasaki lay in ruins, including the Mitsubishi factory where the torpedoes used on Pearl Harbor were made.

Enough was enough.

Emperor Hirohito gave a recorded radio address on 15 August. Never before had the Japanese people heard the shrill voice of their emperor, whom they regarded as the Son of God. 'The war situation,' he said circumspectly, 'has developed not necessarily to Japan's advantage.'[21] Moments later, he announced the surrender of the Empire of Japan.

And that's how the world learned what nuclear power is – an incredibly potent force, hidden for centuries in the tiniest particle of elements deep underground. Scientists discovered this power out of curiosity and developed it further – not to supply energy to society, but to create a brutal military weapon for the world's most powerful country. What nature had hidden so well, they had uncovered. And when they unleashed that power, an explosion of blistering heat occurred. In just a single blast, an entire city was wiped out, leaving tens of thousands of people dead.

Hiroshima became the first symbol of nuclear power – a symbol of total destruction.

CHAPTER TWO

A Welcome Distraction

Why the enthusiasm about nuclear energy?

'It is not too much to expect that our children will enjoy in their homes electrical energy too cheap to meter, will know of great periodic regional famines in the world only as matters of history, will travel effortlessly over the seas and under them and through the air with a minimum of dangers and at great speeds, and will experience a life span longer than ours ... This is the forecast for an age of peace.'

– Lewis Strauss, chairman of the
US Atomic Energy Commission, speech, 1954

In my home country of the Netherlands, newspapers wrote of it with awe: *Het Atoom* (The Atom), a fancy exhibition held at Schiphol Airport near Amsterdam on the energy of the future. In the summer of 1957, after months of hype, the Queen and her husband conducted the opening with something that was supposed to represent a magic wand. After all, nuclear energy is modern sorcery.

At the exhibition, interactive panels provided information and exciting gadgets demonstrated the comforts of modern life. A tray of lights and ping-pong balls depicted the interaction between protons, neutrons and the atomic nucleus. At the entrance, visitors encountered a giant wall made up of 2,500kg (2.5 tons) of coal. They learned that in all this coal there was the same amount of energy as in a tiny metal cap containing just 1g (0.035oz) of uranium.

It was a miracle, indeed.

The highlight was a real nuclear reactor. Crowds queued for an hour before climbing stairs for a view of the apparatus, bathed in a 7m-deep (23ft) water basin. They were enchanted by the mysterious blue glow. The country's largest newspaper spoke of a 'fascinating exhibition'.[1] In just three months, it attracted as many as 750,000 visitors.

This travelling exhibition, which inspired people from all over Europe and as far as India and Japan, was part of a US-sponsored campaign in the 1950s to prepare minds around the world for something as amazing as the energy produced in a nuclear plant. As a Dutch newsreel proclaimed: 'The gradual decline in world supplies of coal and oil makes it necessary for the sake of posterity to develop a new source of energy: atomic energy.'[2]

For uranium to work in a typical power plant, it has to be enriched, so that the proportion of the fissile isotope U-235 increases – not to 90 per cent as in a bomb, but to about 5 per cent. A bomb is designed so that all the concentrated energy is released in a fraction of a second. By contrast, in a nuclear plant, the process of splitting atomic nuclei is regulated by a moderator, control rods and cooling water so that the chain reaction is slowed and stretched, usually for months and sometimes years, before it's time for a fuel change. The heat produces steam that is passed through a turbine and spins the generator. Lo and behold, *electricity*!

Today we think nothing of it, although for some the hint of mystery has never quite disappeared. But in the 1950s, grasping this method of energy production required quite a bit of imagination. Even in the most affluent countries, not everyone had electricity. Power was made possible by generators that could be triggered in various ways, such as by water crashing downwards or by high-pressure steam generated by burning coal or natural gas.

Well, all that was child's play, really. Now, it was possible to obtain energy by splitting the nucleus of the atoms of a tiny chunk of enriched uranium. This was a completely different matter. *This was the future!*

France's first nuclear plant, built in Marcoule, opened its doors in 1956 and became a tourist attraction. Those who had ever been in a coal mine, and tried to brush away the black smudges, marvelled at such a clean facility.

The Dutch government recognised the potential. While some households still burned dried peat, and coal was running out in the mines, demand for electricity was increasing fast. The government committed itself to 'the development of nuclear energy and other economic applications of nuclear fission'.[3] A policy paper estimated that nuclear plants could supply a third of all electricity within 20 years.[4]

However, the plans ended up in the bin when, in 1959, a discovery was made that would propel the Netherlands into modernity. Underneath a farmer's sugar beet field in the province of Groningen, there happened to be one of the world's largest natural gas fields.

The exhibition at Schiphol Airport in 1957 was part of a major charm offensive designed to highlight more peaceful applications of nuclear power. After all, it was clear that atomic bombs had not made the world a more pleasant place. Having also cracked the code of nuclear energy, the Russians had held their first atomic test in 1949. With the rapid expansion of the 'red menace' all the way from Berlin to Beijing, tension flared up between the communist Soviet Union and the capitalist United States. Both superpowers now possessed a superweapon shrouded in secrecy.

During the Cold War, things got heated. In 1952, the US military detonated a hydrogen bomb for the first time – an atomic bomb that gets its explosive energy not so much from nuclear fission as from nuclear fusion. This bomb was a thousand times more powerful than the one dropped on Hiroshima. The blast turned an uninhabited island into a deep crater. Within a year, the Russians had also tested just such a bomb.

These were anxious times. Never before had the human species possessed a weapon with which it could destroy itself. If a government leader were to deploy such a weapon and start a nuclear war, smoke and dust would rise from the burning cities and clump together in the atmosphere. It would block sunlight and no more rain would fall. Earth's temperature would drop to the level of the last ice age. Harvests would fail. For the survivors, famine would be inevitable.

As commander-in-chief, Dwight D. Eisenhower had led the invasion of Normandy, France on D-Day, 6 June 1944. He had seen more than enough suffering. The memory of World War II, and the threat of World War III, were never far from his mind. Now, upon taking office as president of the United States in 1953, Eisenhower worried about society being paralysed by fear of total annihilation. He wanted to provide hope. But how?

His advisers did not quite understand his concerns. They urged additional defence spending for more and bigger weapons. Eisenhower didn't want an arms race, he told them, although he wasn't sure what he wanted instead. So, he requested speaking time at the United Nations General Assembly on 8 December 1953 to address the world, and set about coming up with a plan.

The date was fast approaching.

Then, suddenly, he knew what he would say. Endlessly, Eisenhower polished his speech. It would be a delicate balancing act. He would have to talk explicitly about the dangers and the fears of nuclear power, but he didn't want to scare the hell out of everyone. As his plane approached New York on the day of the speech, he asked the pilot to keep circling above the city; he was still working on the text. Finally, on stage, Eisenhower read from sheets of paper cluttered with deletions, corrections and additions. The speech would go on to become a key moment in history.

First, Eisenhower praised the United Nations. 'Never before in history', he said solemnly, 'has so much hope for so many people been gathered together in a single organisation.'[5] But, he warned, the great tests were yet to come. 'I feel impelled to speak today in a language that in a sense is new,' he continued, 'one which I, who have spent so much of my life in the military profession, would have preferred never to use. That new language is the language of atomic warfare.'

Eisenhower told of the explosive power contained within the US stockpile of nuclear weapons, which was already many times greater than 'the total of all bombs and all shells that came from every plane and every gun' in World War II. How to make such weapons was no longer a secret. This represented a 'danger shared by all'. Now, 'if hope exists', the US president said, 'the hope should be shared by all.'

Then, Eisenhower showed his willingness to phase out the weapons arsenal. Indeed, he wanted to take nuclear technology out of the hands of the military and make it available to people who could use it for peaceful purposes, such as medicine, agriculture and, especially, energy supply.

His vision: the atom should no longer contribute to battle, but to energy.

His plan: the United States would share its knowledge with other countries. Everything from nuclear reactor construction plans to technical assistance would be available, provided the interested countries were open to supervision by a new UN agency to ensure they did not make weapons.

Eisenhower announced that his country pledged 'to help solve the fearful atomic dilemma – to devote its entire heart and mind to finding the way by which the miraculous inventiveness of man shall not be dedicated to his death, but consecrated to his life.'

Applause filled the venue. It continued for several minutes. Even the Russians joined in the clapping. It looked like Eisenhower's eyes were moist.

Next day, the speech was printed in newspapers around the world and commentators were relieved, even laudatory. Many years later, historians would be more moderate in their assessment. They went on to judge the speech, known as 'Atoms for Peace', as a cunning move in international diplomacy.

Eisenhower's promise of disarmament strengthened the Americans' moral position, and with his promise of providing assistance in energy production, especially to developing nations, he wanted to prevent more countries from falling to communism. The aid these nations received was to translate into eternal gratitude to their capitalist benefactor. Thus, the United States worked to build a network of allies.

While the atomic bomb was developed to win World War II, the nuclear plant was used to win the Cold War.

In the wake of Eisenhower's speech, brochures, posters, videos, comics and children's books were produced to spread the word. Stamps featuring the Atoms for Peace logo promoted the policy. PR agents bombarded journalists with press releases and proposed articles on the wonders of nuclear power. They showed that the dreams of the future went beyond electricity. Nuclear technology would also be useful in medicine, for disease detection and treatment. In agriculture, radioisotopes could be used to protect harvests from spoilage. In countries facing drought, nuclear reactors could provide the energy for desalination, so there would never again be conflicts over scarce water. They would also provide power for large-scale fertiliser production, so that fields would always be fertile. Moreover, nuclear energy meant a revolution in transport, providing fuel for trains, ships, planes and rockets.

The enthusiasm was contagious. In the Netherlands, a leading newspaper proclaimed in 1954: 'Nuclear power can prevent millions from perishing in the year 2000'. Subheading: 'There's still enough fuel for several centuries'.[6] Inspired, the KLM Royal Dutch Airlines director boasted that the airline would soon have several aircraft flying on nuclear power and talked of a 'super-network of long-distance aircraft with atomic power'.[7] The Dutch government collaborated with one of the wealthiest families to 'develop the application of nuclear energy in the interests of Dutch maritime shipping'.[8]

One observer foresaw the nuclear-powered car but drew the line at a cigar lighter with an 'atomic pocket battery'.[9]

Oh yes: nuclear power, it was being said everywhere, was the solution to everything.

In his television series *Disneyland* in 1957, Walt Disney devoted an episode to nuclear energy, showing 'our friend the atom' as a willing servant, but one with pent-up anger – so much, in fact, that the atom has an unimaginably destructive side.[10] Fortunately, scientists knew how to contain this constant threat.

The enthusiasm was equally alive in the Soviet Union. According to party officials, the atom heralded a communist utopia. Nuclear plants would bring heat and light to people's homes everywhere. The technology was praised in newspapers, on the radio, on the cinema screen and in primary schools.

The Russians took pride in being the real pioneers. As far back as 1951, they had started building the very first nuclear plant in Obninsk, 100km (60 miles) from Moscow. Three years later, it was connected to the power grid. Nikita Khrushchev, successor to Joseph Stalin, liked to tell everyone that this was yet more proof of socialism's superiority. The name of the modest nuclear plant: *Atom Mirny*, or Peaceful Atom – a subtle sneer at American militarism.

Other countries followed suit. Once at the forefront of coal, the UK now wanted to be at the vanguard of nuclear power. Canada believed nuclear energy could start a revolution just as the steam engine once did, and was building on the knowledge it gained as a partner in the Manhattan Project. France was looking to restore its old glory in the modern, postcolonial world, embracing nuclear technology as a prestige project that belonged to the global elite.

In the early days, many nuclear reactors used graphite as a moderator to slow down neutrons and split U-235 nuclei. For coolants, they worked with water, gas or sodium. Others used heavy water (deuterium oxide) as both moderator and cooling agent. The winning, most widely used model today is – you guessed it – American-made. In the pressurised water reactor, a method originally developed as a compact propulsion system for submarines, water is pumped under high pressure to the reactor core where it is heated and turned into steam to drive a turbine. Water functions as both moderator and coolant.

Eisenhower got his way when, in 1957, the International Atomic Energy Agency (IAEA) was established within the United Nations to inhibit the use of nuclear energy for military purposes. By then, the Netherlands was among 50 countries with which the Americans had signed a contract for the construction of reactors or the supply of enriched uranium. Leading up to the exhibition at Schiphol Airport, the US sent enriched uranium along by plane. It was packed in a 100-kilo (220lb) wooden box, lined with lead to protect anyone who picked it up from radiation. 'Caution', it said on the side.

The advance of nuclear power proceeded fast. Its heyday was in the 1960s, when it was still cheap to build nuclear plants and they were getting bigger and bigger. Never before had a new energy source taken such a large market share so quickly.

The optimism about nuclear power was welcomed. After all, the future of energy supply looked bleak. In large parts of Europe, the best coal mines were pretty much depleted and wherever there could be a hydroelectric plant, there already was one. Now that Saudi Arabia appeared to have large oil fields, oil was presented as a potential source of electricity. But could we rely on it? Egypt closed the Suez Canal in 1956, and suddenly the supply of oil stagnated. Some countries introduced a car-free Sunday – one sign of an oil crisis. It would be just the first.

In the Netherlands, experts gathered to deliberate on the rising energy demands of a growing and more prosperous population.[11] Where should the energy come from, was the main question of a 1957 conference in Amsterdam. Should it come from the sun? Alas, that would require a third of the

Dutch land surface to be covered with 'solar mirrors', experts concluded. Should it come from the wind? No, was the verdict. Even 'if we covered our entire coastline with a continuous row of windmills and managed to convert all the energy from the wind into useful energy for us – which is completely impossible anyway – we might just have enough to cover a quarter of our present-day consumption of electrical energy.'

Other concerns came from a recent news article with the headline: 'Earth is becoming more and more of a greenhouse'.[12] The first sentence read: 'The burning of fossil fuels such as coal and petroleum and the resulting increase in carbon dioxide levels in the Earth's atmosphere may have very unpleasant consequences for the global climate in about 50 years' time.'

All in all, there was clearly a problem.

But, thank God, there was nuclear power. With nuclear reactors, a country could be a lot less dependent on coal, oil and gas. And should international conflicts arise close to home, knowledge of nuclear technology might come in useful.

This realisation motivated European integration. To ensure nuclear energy was used for peaceful applications, oversight was required and cooperation was needed. On 25 March 1957, six countries signed a treaty to establish the European Atomic Energy Community, aka Euratom. That same day, they set up another body: the European Economic Community (EEC). The signatures of the six heads of government formed the basis of what would later become the European Union. As Jean Monnet, a cognac merchant and civil servant now remembered as 'the father of Europe', wrote in his memoirs: 'The peaceful atom would become the spearhead for the unification of Europe.'[13]

Nuclear energy united an uncertain continent, spurred on by that self-assured country on the other side of the ocean, which, despite bulging with fossil reserves itself, was pioneering a completely new energy source.

However, while public opinion on nuclear power was becoming more positive, not everyone joined the choir. Critics were sceptical of the utopian dreams. Others doubted whether scientists and engineers were worthy of so much public trust. For some, nuclear fission was hubris.

A Dutch local newspaper decided to have some fun with all the rosy talk of a nuclear future. A seemingly earnest article spoke of a unique experiment with 'atomic paste', enriched uranium one could squeeze out of a tube. As a world first, 10 buses in the city of Tilburg would run on just a pinhead of this blue lubricant, which could easily be applied near the engine block. The trial run, on a deserted road after midnight, got off to a flying start. 'After only a few metres', wrote the reporter, 'the bus had almost crashed at dizzying speed against a tree.' Surely everyone wanted to experience such novelty? To avoid disappointment, the newspaper advised its readers to ask the bus driver before boarding: 'Are you running on atomic paste?'[14]

Two days later, the newspaper disclosed that it had been an April Fool's Day joke.[15]

For a long time, uranium's secrets remained hidden. It was made billions of years ago in exploding stars. When our galaxy formed from the dust cloud, uranium entered the planets. On ours, under the Earth's crust, uranium would provide its new residence with heat – as well as the power to destroy it. Looked at in this way, according to Tom Zoellner in his book *Uranium*, Mother Earth has been burdened with 'a geologic original sin' from its inception.[16]

Archaeologists have shown that people living in ancient times had already dug up uranium. The Romans used it to colour their stained glass. Indian people put it in the paint they used on themselves for religious occasions. Other than that, the rock had no value.

At the end of the Middle Ages, uranium surfaced in large quantities from a silver mine in Bohemia, where the border between the Czech Republic and Germany now lies. Here, in a town that became known as Joachimsthal (now Jáchymov), blacksmiths minted large, heavy silver coins called *Thalers*, the word from which 'dollar' is derived. Meanwhile, the miners had stumbled upon a dark, greasy stone. It stuck to their picks in a blob. They threw it away.

Some wondered what this stuff was and why the rocks next to it tended to have such a green or orange colour. No one knew. They *did* know, however, that miners got sick. After years of working in the silver mine, they developed a persistent cough and spat blood. Once their lungs had slowly rotted away, death wasn't far off. Doctors were puzzled. This was nothing like the plague or other diseases they knew in the sixteenth century. Was it perhaps due to the air in the mine shafts? They didn't know for sure. They called it *Bergsucht*, mountain sickness.

It was not until 1789, the year the French overthrew their king and the Americans elected their first president, that uranium was officially identified as a chemical element. The achievement came in the name of Martin Heinrich Klaproth, a German chemist. In his laboratory in Berlin, Klaproth liked to carry out meticulous research. He dissolved the unknown substance in nitric acid, added potassium salt to the yellow sediment, mixed it with wax, added some oil – easy, right? The chemical reaction that followed was unlike any other. Klaproth had stumbled upon something new.

In naming his element, Klaproth was referring to a planet recently spotted by a composer with a home-made telescope.[17] It was the first planet to have been discovered for a long time, and the first one outside the then-known boundaries of the solar system. News of Uranus led to a revival of interest in astronomy.

A century and a half later, the focus was no longer on stars and planets moving around the sun. Far more interesting were the electrons and neutrons moving around the atomic nucleus.

By then, the early nuclear weapons designers thought uranium was rare. Just about all of it came from a country with an extremely violent past. The Democratic Republic of Congo had its origins in 1885 as an enterprise owned by Leopold II, King of Belgium. The monarch was keen to join the European inclination to rule the African continent, extracting and trading its raw materials. Leopold II turned the 'Congo Free State', with its unprecedented wealth of mineral resources, such as copper, rubber, gold, tin, zinc, cobalt and diamonds, into a permanent labour camp. If the locals didn't work hard enough, they were whipped. Those who didn't listen had their hands cut off, or else their heads.

In 1915 – by which time Leopold had handed over his private colony to the Belgian government – a British prospector in the region stumbled upon uranium ore.[18] While uranium seemed pretty useless, it hid something that was at the time more valuable than gold: radium, which in certain circles was conceived as a health-promoting panacea. In Shinkolobwe, a mine in the south of the country, the ore proved to be much purer than anywhere else. It was dug up and sent to the port for a one-way trip to Antwerp.

By the time physicists were preparing for nuclear fission and government leaders were preparing for war, the mine in Shinkolobwe was already closed. But Belgian Congo, known as the main supplier of uranium, still had leftover supplies. Military strategists got in touch. Hence the explosive material in the atomic bombs that American soldiers dropped on Asian civilians came from African workers and was traded by European businessmen.

After the war, the US government wanted all the world's uranium, especially once they learned the Russians were producing, even testing, atomic bombs as well. Whoever found uranium was paid handsomely. In their hunt for the new gold, adventurers bought a Geiger counter, the radiation detector named after Hans Geiger, a former colleague of Ernest Rutherford who worked on developing the German atomic bomb. Some of these prospectors made a fortune. One downtrodden geologist became a millionaire after a spectacular find and bought a villa with a swimming pool and accommodation for servants.[19] The uranium rush led to several places proudly calling themselves the Uranium Capital of the World. In Grants, a small town in New Mexico, the Uranium Café served Uranium Burgers. There was no uranium in them.

In the years following the war, the atomic bomb was seen as exciting, almost seductive. In cities all over America, nightclub revellers could enjoy performances by 'atom bomb dancers'. Next, they might order an atomic cocktail: a mix of, say, rum and blue curaçao with added dry ice, so that the drink bubbled and smoked. The winner of a beauty contest even crowned herself Miss Atomic Bomb while dressed in a fluffy white dress the shape of a mushroom cloud.

Elsewhere, Louis Réard, a French automotive engineer who had taken over his mother's lingerie business, introduced a two-piece swimsuit in the summer of 1946. He based his creation on a rival's design that the French called 'l'atome', which was touted as the smallest swimsuit in the world. Réard named his new, even smaller version after the little island in the Pacific Ocean where the Americans had just tested an atomic bomb: Bikini. It was so revealing that no model wanted to wear it, upon which an 18-year-old nude dancer, one Micheline Bernardini, was hired. The Paris fashion press struggled to make sense of the name 'bikini'. A reporter suggested such small patches of clothing would be all that was left after a nuclear blast.

Visit Las Vegas at this time and you could also try your luck outside the casinos. With help from the local tourist office, you could book a bus trip to a spot with stunning views of a nuclear test in Nevada's vast desert. School classes even went there on field trips. Put on those sunglasses and enjoy! Hotel rooms in Las Vegas with a view of the blast charged a premium. For those who missed it, the film version could be seen on television and in cinemas, sometimes with commentary by a star actor.

An atomic test served many purposes: to test a new design, to investigate basic physics, to make sure an ageing warhead still worked, to impress the enemy – there was always a reason. The Americans performed their tests in sparsely populated states like Nevada, New Mexico and Alaska, or on islands in the Pacific Ocean. The French did it in Algeria and French Polynesia, the British in Aboriginal territory in Australia.

The tests went on and on. In just three months in 1961, the Russians detonated 50 atomic bombs with an explosive power greater than that of all previous atomic tests combined. The heaviest of all, the *Tsar Bomba*, resulted in a mushroom cloud 60km (almost 40 miles) high over Nova Zembla. In Finland, almost 1,000km (over 600 miles) away, windows cracked.

Sometimes the consequences were more far-reaching than damage to glass. Another atomic test, on Bikini Atoll in 1954, went awry. The wind, in the legendary words of the authorities, 'failed to follow the predictions', and the bomb scattered radioactive fallout over the crew of a tuna fishing boat.[20] The 23 men aboard the *Daigo Fukuryū Maru* (Lucky Dragon No. 5) were sprinkled with a drizzle of curious powder. Astonished, they rubbed it in their hands. One of them licked it. It went on falling for hours. Before long, they were suffering from headaches, dizziness and diarrhoea. Their hair fell out and their skin was covered in blisters.

At the hospital in Tokyo, where one crew member died of complications from a blood transfusion and the others stayed for more than a year for painful but ultimately successful treatment, the doctors were familiar with these symptoms: they knew them from the survivors in Hiroshima and Nagasaki.*

---

* The fishing boat incident became the inspiration for the film *Godzilla*, about a prehistoric sea monster awakened from sleep by an atomic test and set on revenge.

While American schoolchildren were taught through cartoons to crawl under the desk at the sound of an air raid alarm, and their parents were assured in leaflets that the radioactive fallout from atomic tests was innocent, clumsy nuclear weapons testing continued, usually followed by denials and lies from the authorities.

In the Netherlands, the opening of the Schiphol Airport exhibition was covered by *De Telegraaf* on the very same page as a news article about a failed nuclear test in the Nevada desert. An 'A-bomb' had not exploded as planned, and five technicians 'with sweat on their foreheads' had to climb a 150m (490ft) tower to defuse the thing.[21]

Slowly, initial enthusiasm for 'atoms for peace' was tempered with a sobering awareness of the dangers of nuclear technology. Anxiety increased everywhere. The Dutch were starting to feel uncomfortable about US nuclear weapons being stationed in their country. In 1961, a first protest against nuclear weapons was held in Amsterdam. In a survey, three in four people in the Netherlands estimated that the chance of surviving a nuclear war would be 'small' or even 'nil'.[22]

During this time of wavering peace, increasing numbers of people realised that nuclear plants were more closely related to nuclear weapons than authorities had suggested. For instance, colourful brochures documenting the miracles taking place inside the Calder Hall nuclear plant in Sellafield, Cumberland, UK, opened by a 30-year-old Queen Elizabeth II in 1956, were handed out to the people cheering along her route. In defiance of the pretty talk and shiny pictures, the plant did not so much provide power for the grid, but rather mainly plutonium for British atomic bombs.

If you want to produce nuclear weapons, you enrich uranium further and further, as occurred with the bomb on Hiroshima, or you shoot neutrons at depleted uranium in a nuclear reactor until you get plutonium, the material used in the bomb on Nagasaki. In fact, if you want to produce electricity in a nuclear reactor, you will also get some plutonium as a waste product. This was enough for people to suspect, even in operations focused on power production, that it was all about the bomb.

To demonstrate the obvious link, they only needed to point to the Atomic Energy Commission (AEC) in the United States, which oversaw both nuclear plants and nuclear weapons. The basic physics of the two were the same and the same people were involved. Yet, one important difference was overlooked by some: if you think of a nuclear weapon as an outsized stick of dynamite, with the energy bursting out all at once, a nuclear plant is a mere stick of incense, with the energy being calmly released.

When imaginations run wild, however, thoughts stray to the mad professor. For decades, he had been a popular character in the science fiction genre – a brilliant scientist, misunderstood and unloved, driven by honour and revenge. From a darkly lit laboratory, he worked on a powerful weapon that would bring disaster across the world. Could such a mad professor secretly create a nuclear weapon?

There did not even have to be malicious intent. A simple mistake was enough, it was thought. And mistakes there were. More than once, nuclear weapons fell accidentally from patrolling bombers. Thanks to safety systems, explosions were prevented, but there was also the occasional close call when early warning radar systems appeared to signal that the enemy had launched a nuclear attack.

Atoms for Peace may have led to a boom in nuclear plants, but it was overshadowed by the proliferation of nuclear weapons. When Eisenhower stepped down as US president after eight years, the number of nuclear weapons in the world had risen from just over 800 to nearly 20,000.

A simple logic settled in: more nuclear power leads to more nuclear weapons. After all, the owners of these bombs seemed to have built the very industry needed to produce them. Open the door to nuclear plants, and it's only a small step to all-out war.

A poignant illustration of the delicate link between applications for war and peace came in the person of Abdul Qadeer Khan from Pakistan. After studying in Germany, Belgium and the Netherlands, Khan landed a job in 1972 at an engineering firm in Amsterdam. Here, he performed research on some of the materials used by the British–German–Dutch consortium Urenco, where ultracentrifuges are used to enrich uranium. He was soon offered a senior technical position at the organisation.

Khan's work involved classified information, accessible only to authorised personnel. But his curiosity was insatiable. One colleague gave in and took him to the rooms Khan was not supposed to enter. The colleague also gave him detailed photographs of equipment, which he later saw lying around at Khan's home. Their employer only sensed something wasn't right in 1975, when Khan had shown a little too much interest in parts of an ultracentrifuge at an international trade fair. He was transferred promptly. That year, Khan and his wife left for their annual family visit to Pakistan, and didn't return.

Instead, Khan joined Pakistan's secret programme for an atomic bomb and opened a laboratory in 1976. With the knowledge he had gained at Urenco's facility producing fuel for nuclear plants, Khan became a dealer in blueprints for nuclear weapons. He sold them to regimes in Iran, Libya and North Korea. In 1998, Pakistan tested its first nuclear weapon, and Khan became known in his homeland as 'the father of the Islamic bomb'.

Even without the threat of a future nuclear jihad fighter, there was clearly uncertainty from the get-go as to whether nuclear power was being used for peace or for war, or both, and whether that distinction was clear to everyone working with it. Some observers decided that a nuclear plant was simply a bomb incubator disguised as a power facility.

Indeed, one could be forgiven for thinking that the nuclear power industry was conceived for one purpose: to divert attention from nuclear weapons.

CHAPTER THREE

The Reckoning

Why the resistance to nuclear energy?

'The great enemy of the truth is very often not the lie, deliberate, contrived and dishonest, but the myth, persistent, persuasive and unrealistic.'

— John F. Kennedy, US president, 1961–1963,
speech at Yale University, 11 June 1962

Most of those who showed up were well prepared for the freezing cold, but not for a battle. On the morning of 28 February 1981, tens of thousands of men and women, young and old, set off for Brokdorf, a village near Hamburg. They defied a demonstration ban and evaded the roadblocks. They came as construction of a nuclear plant, halted for years due to protests, was finally about to resume.

The protesters were concerned that a nuclear reactor could not be contained. A malfunction or mistake could release radiation into the air. They wondered if the waste could ever be disposed of safely. They believed nuclear power was a risky operation that was simply unnecessary. After all, wasn't there solar energy? And couldn't we manage with less energy overall?

Their slogan: *Atomkraft? Nein danke!* (Nuclear power? No, thanks!)

Most of the protesters didn't go near the construction site. It was fenced off and surrounded by meadows and frozen ditches, but some wanted to get closer. That's when the mood changed. A police cordon awaited them; helicopters circled overhead. Members of a militant group began launching stones and Molotov cocktails. Police officers responded with tear gas and water cannons. When, during a chase, one of the officers tripped and fell, two activists rushed in and beat him with a baton. By the end of the day there were wounded on both sides, as many as about 200 altogether.

The clash was an illustration of the tense mood across Western Europe. In Italy, former Prime Minister Aldo Moro had been kidnapped and murdered by the Brigate Rosse, the Red Brigades. In Germany, the *Rote Armee Fraktion* (Red Army Faction) had done the same with industrial leaders. In England, there were attacks by the Irish Republican Army, and in Spain, by the ETA (*Euskadi Ta Askatasuna*) separatist group.

By the time the German protesters gathered in Brokdorf, European actions against nuclear power had already resulted in one death. In 1977, at the Superphénix nuclear plant, under construction by the Rhône River between Lyon and Geneva, a protester died after a police crackdown on a gathering of 60,000 French activists. Five years later, one of them returned to the site. Fearing that a nuclear plant could explode like an atomic bomb, he, perversely, fired five rocket-propelled grenades at the building. Two of them hit the plant. The reinforced concrete wall was slightly damaged, though the construction workers remained unharmed.

At the time, there was no trace of the perpetrator. But more than 20 years later, after the case was time-barred, Chaïm Nissim, son of a Jewish banker in Switzerland, confessed to firing the grenades. He called it a 'non-violent action'.[1] In his book, *L'amour et le monstre*, he revealed that the grenade launcher was supplied by Ilich Ramírez Sánchez, also known as Carlos the Jackal, then one of the world's most wanted terrorists.

In the nuclear reactor, neutrons split atoms; outside, they split society.

The European movement against nuclear power had peaceful beginnings when it was formed in the early 1970s, with silent protest marches and citizen-led petitions. The cradle of the movement is usually placed in Wyhl am Kaiserstuhl in the Upper Rhine Plain, Germany. For the surrounding region, a grand plan had unfolded in 1970: a number of nuclear plants were to provide energy for industry around cities such as Frankfurt, Strasbourg and Basel. The small town of Wyhl was included in the plan.

At the time these plans were drawn up, most people in Wyhl and elsewhere in Western Europe didn't know much about nuclear energy. Some remembered a pompous exhibition and some knew of a research reactor here and there, courtesy of the Americans. But beyond that? Germany, Switzerland, Spain, Sweden and the Netherlands still had only one nuclear plant each connected to the grid. Italy had three. The two nuclear-armed superpowers, France and the UK, had around 10 each in operation. Belgium, Austria and Finland were yet to start working on them.

A 1958 novel had tried to make Europeans familiar with the technology. Günther Schwab, an Austrian writer and former forest ranger, became an influential player in the intellectual vanguard of the struggle against nuclear power thanks to *Der Tanz mit dem Teufel* (the English translation was published under the title *Dance with the Devil*).

Schwab outlines a secret demonic society that updates the 'Boss' Devil on devious attempts to exterminate humanity. His devil employees report in detail about how they're destroying the Earth's habitat. They have devised all sorts of evil gimmicks, from new pathogens and artificial food additives to incessant music from radio sets. Their worst weapon of mass destruction is saved for one of the final chapters, when a devil named Stiff appears on the scene. He's the chief of the department of 'atomic death'.[2]

His face is expressionless, Schwab writes, 'sleepy yet full of restrained energy'. Sixty years before, Stiff had been appointed to 'find some new power of destruction'. He was there, looking 'over the shoulder of those madmen who, divorced as they were from life, worked out the great secrets of atomic science'.

The devil then sneakily directed the work of physicists so that minute amounts of invisible radioactive substances would poison the soil, the water and the air, killing plants, animals and mankind. At first the radiation came from atomic bombs, and later from the cover of nuclear plants, Stiff explains. The 'atomic poison' is released when the waste bursts out of the storage containers. Also, when something goes wrong at a nuclear plant, radioactive material will leak and nearby residents will be poisoned without anyone noticing. Stiff smiles. Just the vapour rising from such a reactor could spell disaster! When the 'Boss' Devil has finished listening to Stiff's account on the prospects of atomic death, he says: 'Excellent!'

One reviewer of Schwab's book wrote it was an 'alarm bell for modern society'.[3]

Along with the book, Günther Schwab launched an organisation that advocated for nature conservation: *Weltbund zum Schutz des Lebens* (WSL), or World Union for Protection of Life. Dozens of Nobel laureates sat on its scientific advisory board. The legendary American chemist Linus Pauling served as its president. With their support, Schwab was leading the fight against nuclear energy, and as early as the 1960s, WSL led a protest against the construction of a nuclear plant in Würgassen, near Göttingen.

Historian Joachim Radkau would refer to the German WSL chapter as the 'nucleus' of the anti-nuclear movement and Schwab as a pioneer who gave 'a significant impetus' to the campaign.[4] In the mid-1970s, Schwab's book on the devils was touted in the press as a 'standard work on environmental protection'.[5]

However, the reputation of the author and his organisation was questionable. In the 1930s, Schwab had been a member of the Nazi Party and joined the *Sturmabteilung* (SA), the paramilitary wing of Hitler's political party that stormed Jewish shops and attacked Jews in the streets. Schwab had climbed to the rank of *Sturmführer*. After fighting on the Eastern Front during the war, he'd published a magazine that was regularly accused of scientific racism. Throughout his life, he remained an outspoken supporter of population control and eugenics, lamenting the 'loss of prestige of the white race'.[6] When Schwab died in 2006 with a major Austrian award for the arts and sciences to his name, his legacy included a vast collection of National Socialist writings.

In Germany, Schwab's WSL was led by kindred spirits. One chair was a clergyman who in the 1960s founded an eco-fascist organisation that would ultimately be banned for 'continued denial of the Holocaust'.[7] Another chair was a naturopath and author of popular books on natural healing who would be identified by an investigative journalist as the 'bridge between the ecological movement and neo-Nazis'.[8] Yet another made international headlines in 2015 when she was sentenced to prison for Holocaust denial at 87 years of age.[9]

Meanwhile, the winegrowers around Wyhl am Kaiserstuhl didn't seem to share Schwab's concerns about radiation, accidents or waste. They feared not so much the supposed wider damage to nature, but the damage to their grapes. As Stephen Milder recounts in his book about the rise of the anti-nuclear movement in Germany, *Greening Democracy*, the people of Wyhl weren't particularly fond of the concrete tower, where unused heat from the reactor would be cooled by water. The vintners were convinced that the water vapour would lead to fog lingering in the valley. If that were to happen, the sun would be blocked out and their chic wine would turn into some ordinary table variety, spelling financial disaster. No one, not even a meteorologist, could allay their concerns.

In Freiburg, the nearby university town, anti-nuclear posters were hung showing the city's slender cathedral spire contrasting starkly with an immense cooling tower.

Uproar there may have been, but in a referendum, most Wyhl residents voted in favour of selling the site to the nuclear plant operator. Some hoped it would make the community a bit of money, others hoped for jobs. The worries about the fog were apparently not as widely shared as first appeared. A similar response could also be seen across the French border near Fessenheim, where locals were annoyed by protest marches organised in their village by opponents of nuclear power.

And then, on 17 February 1975, a month after the referendum, when anxious citizens were still talking with municipal politicians, an entire area at Wyhl was cordoned off. Without notice, construction began.

Some locals were not amused. They called a press conference a day later and expressed their displeasure. One journalist asked whether there would be an occupation of the site. After all, there had been one in Marckolsheim, France, across the Rhine, hadn't there? The locals shook their heads. No, there would be no occupation in Wyhl.

After the press conference, a group of people – mostly women – decided to walk towards the construction site. They just wanted to have a look. At the site, they saw men with bulldozers, men with excavators, men with chainsaws. For some, the spectacle was too much. They demanded the workers' attention. They shouted and lashed out at the fences, managing to knock one down. With no real plan, they walked onto the site. Over here, someone climbed on the bucket of an excavator; over there, a group huddled around a bulldozer. It was chaos. Within an hour, the work stopped, but the visitors did not leave. It would be an occupation after all.

Wyhl am Kaiserstuhl became national news thanks to the state's minister-president, who suggested that the protesters were enemies of democracy. In a country still coloured by its Nazi past, this was a serious accusation. Six hundred police officers sprang into action. They came in armoured cars, bringing water cannons and a pack of dogs. That is how they chased away the 150 villagers remaining at the construction site. More than 50 were arrested.

The police crackdown in Wyhl was strongly condemned, including by those who saw the protest against the nuclear plant as a convulsion of narrow-minded provincialism. Some opponents of the nuclear plant, and of all nuclear plants everywhere, sought revenge. For with the clearing of the construction site in Wyhl, the opposition to nuclear power had not been brushed away. A few days after the evacuation, trained activists cut a hole in the fence and began an epic occupation that would last nine months.

The events at Wyhl were the definitive breakthrough of a social movement in Europe that would work to thwart nuclear power by any means necessary. Germany was at the vanguard.

The nuclear plant at Wyhl never came to be. The area is now a nature reserve.

In the United States, a movement against nuclear power had been around for some time. Again, the initial concerns had nothing to do with the environment but rather worries that nuclear weapons would be fired and the world would end. With the Russians, far away in the Kremlin, one could never know, but the men in the White House couldn't be trusted either. The fear was firmly entrenched.

Atomic tests were necessary and perfectly safe, the US government said. Not everyone agreed. Scientists issued warnings. Even a minor fallout would affect the food chain and weaken our bodies. Some predicted that exposure to radioactive particles could lead to an epidemic of cancer and genetic mutations. One scientist claimed that already hundreds of thousands of children had died from the radiation released by atmospheric testing.[10] It wasn't true and there was no proof, but it sounded very bad.

The American people realised that while they might not be able to prevent a nuclear war, they might be able to prevent atomic tests. Once they started organising, they could perhaps influence politicians. And so, in 1954, Americans took to the streets for their first large-scale protest seeking to ban atomic tests. The ban came, at least for above-ground tests, because after an international treaty in 1963, the tests continued underground. Fallout was now eliminated. While the nuclear weapons clamour raged on under our feet, even at an accelerated pace, there was now little public protest. Yet concerns – about a possible nuclear war, about possible radiation exposure – would not budge. So what happened? Attention shifted to nuclear plants.[11]

Doesn't a reactor involve radiation too? Doesn't a leakage release radioactivity? Is a nuclear plant not simply a breeding ground for nuclear weapons? Surely such a building could just explode like an atomic bomb, right? Such questions led to a new goal: that of getting nuclear plants banned.

Greenpeace offers a striking illustration of the shifting focus. This organisation grew out of a pacifist Quaker community that campaigned against an underground atomic test near Alaska in 1969. From Canada, Greenpeace's international breakthrough followed in 1973. France, which had not signed the treaty banning above-ground testing, wanted to detonate an atomic bomb near an atoll in the Pacific Ocean. David McTaggart, a former badminton champion and property developer from Canada, sailed his personal yacht provocatively close to the atoll, preventing the atomic test from being carried out.

At sea, French marines attempted to dissuade McTaggart and his fellow travellers, to no avail. When they came aboard *Greenpeace III*, they beat up McTaggart, permanently damaging his right eye. The French government later denied the accusations of violence. However, a Greenpeace crew member had secretly taken photos and called in the press. As the photos circulated around the world, the French government reluctantly announced it would stop above-ground nuclear testing. Greenpeace's name was established.

Later, McTaggart wrote about the deeper motivation of the fellow activists on his sailing boat in his book *Greenpeace III:*

Journey into the Bomb. He singled out two younger men on board whose lives were shaped by the atomic bomb. 'Previous generations had trembled before visions of hellfire and brimstone, but theirs was the first generation for whom the hellfire was real … No wall existed that it could not breach. No hole existed that was deep enough for escape. There was no place on earth the fire could not reach. And no father's arm could hold it at bay.'[12]

When McTaggart wrote these powerful words in 1978, Greenpeace had branched out to around 20 countries. By then, it was staging high-profile direct actions against nuclear power everywhere, with, for example, protests against the dumping of radioactive waste discharge in the ocean. While the threat of nuclear war dwindled into the background, opposition to nuclear plants became a signature part of the organisation.

Today, Greenpeace's websites inform us that nuclear power is 'dirty and dangerous', that it can cause 'disastrous catastrophes', that it creates 'huge amounts of hazardous waste', and that it 'has no place in a safe, clean, sustainable future'.[13] But they contain barely a word about nuclear weapons.

Just as the production of plutonium for nuclear weapons merged with the supply of electricity, or vice versa, so opposition to a nuclear arms race merged with opposition to nuclear plants. Those who agitated against nuclear technology targeted a clique of belligerent, war-mongering governments and reckless, profit-obsessed energy companies.

The *Encyclopaedia Britannica* describes the anti-nuclear movement as a social movement 'opposed to the production of nuclear weapons and the generation of electricity by nuclear power plants'.[14] On Wikipedia, the pages on the movement offer a hodgepodge of criticisms of bombs and reactors.

'No nukes!' soon became a popular slogan. In the 1960s, this had meant 'No nuclear weapons!' One decade later, it could just as well have meant: 'No nuclear energy!'

The mash-up of nuclear weapons and energy even transferred to pop music. In 1979, a concert in Madison Square Garden in New York City, with a line-up including Jackson Browne, Bruce Springsteen and James Taylor, drew 200,000 visitors, who chanted: 'No nukes!'. They wanted to phase out nuclear energy. In 1982, another concert, this one in NYC's Central Park but with a line-up including the exact same headliners, drew a million visitors. Again they chanted: 'No nukes!'. This time they wanted to phase out nuclear weapons.

Just as the producers of nuclear weapons and nuclear energy made use of the same knowledge and experts, the movement against nuclear weapons and nuclear energy involved the same slogans and people.

Following its spectacular entrance onto the world stage, nuclear power had been tamed and gloriously resurrected as a radically new way to deliver energy. While reflecting on his lifelong career in the nuclear field, a former Manhattan Project lab worker noted that he and his colleagues were 'tainted by the bomb'.[15] They all felt better thanks to the nuclear power plant. 'What could be nobler', he wrote, 'than providing the world with an ultimate energy source?'

Well, not everybody appreciated it.

Those who opposed nuclear power challenged the establishment. They distrusted science, politics and big business, and would no longer keep quiet. Since the emancipation wave of the 1960s and 1970s, they wanted to have a say in everything: on capitalism, war and oppression in general. Every opinion counted, especially their own.

When it came to nuclear power, supporters and opponents talked past each other. One group believed nuclear power was clean and safe, while the other thought it dirty and risky. Both sides acted as if they had knowledge, whereas the other side only had ideology. They made the same accusations: their adversaries were thoughtless, superficial, short-sighted. In the public debate, the opponents of nuclear power won. They showed their emotions and appealed to those of others, while nuclear advocates came across as detached accountants.

This clash illustrated a growing gap between the scientific world and literary intellectuals. The divide was identified as early as 1956 by a man who had affinity with both sides: Charles Percy Snow, better known as C.P. Snow. He had studied physics, worked under Ernest Rutherford and served the British government as a scientific adviser. Everyone knew him as a writer of popular novels.

In what is considered to be one of the most influential lectures ever, Snow stated that there was a deep mutual distrust between science and the humanities. Between the two cultures, the relationship was one of hostility. Above all, there was a lack of understanding. Snow was annoyed that the scientists among his friends even spoke of Charles Dickens as if he were an obscure experimental author. Of the writers among his friends, he found it unacceptable that they had about as much insight into modern physics 'as their neolithic ancestors would have had'.[16]

Snow noticed that literary intellectuals were gaining influence in politics. This worried him. It is dangerous, he warned, when policy is increasingly made by people with an interest in the classics but not in the natural sciences, a discipline that had arguably given people richer and healthier lives.

What C.P. Snow did not and could not see was that it was precisely this increasing prosperity that would soon be challenged.

One of the first to question this idea of progress also worked for the British government, at the intersection of politics and industry. He too wrote eloquent commentaries. And he too became known by his initials. E.F. Schumacher was born in Germany, but moved to England when the Nazis were preparing for a war in which his brother-in-law, Werner Heisenberg, was put in charge of developing Hitler's atomic bomb.

In his new homeland, Schumacher, trained as an economist, worked alongside the renowned John Maynard Keynes. Together they reordered the global economy, creating institutions such as the World Bank and the International Monetary Fund, and sparking rapid post-war recovery. In factories, automation increased productivity, while agriculture saw higher yields thanks to the introduction of fertilisers and synthetic pesticides. Economies boomed, unemployment fell and wages rose.

Daily life was becoming more pleasant for everyone, but Schumacher felt a growing unease. Progress, he thought, was breaking the bond with ourselves, with each other, with nature, with the divine. He referred to industrial society as 'the mother of the bomb', as it had grown out of the same roots as the atomic bomb – 'a violent attitude to God's handiwork instead of a reverent one'.[17] He concluded: this is all a great failure.

In lectures and essays from the 1960s onwards, Schumacher asked pressing questions. Is economic growth really such a good thing? Doesn't it contribute to social disruption? How long can growth be sustained? Isn't there more to life than money? Should technology keep advancing? Why should companies, profits, cities, the world population, everything, keep getting bigger and bigger?

Schumacher had answers, too. The worship of economic growth was wrong. Growth had ruined us. He used inverted commas for words like 'modernisation' and 'development' to illustrate his scepticism.

In one of his essays, Schumacher addressed C.P. Snow directly. He deemed the optimism of scientists, which Snow said was needed to solve problems, as 'guilt-ridden' because of the damage their inventions had caused.[18] No scientific or technological advance, nor any political or economic reform, could solve 'the life-and-death-problems of industrial society', he said, because the problems 'lie too deep, in the heart and soul of every one of us. It is there that the main work of reform has to be done – secretly, unobtrusively'.[19]

Time to change course.

Luckily, Schumacher knew the right course. In 1955, he found inspiration in Burma, now Myanmar. He was there for three months to give economic advice, an important part of his role during the 20 years he worked as a top adviser to the British nationalised coal industry. As a representative of one of the world's largest organisations, he defended the interests of coal against the rapid rise of oil and nuclear power – new, competing energy sources, about which he aimed to curb the apparent 'limitless optimism'.[20]

In Burma, Schumacher visited several villages before concluding that the locals needed little advice. Indeed, Western economists could learn something from *them*. 'They have a perfectly good economic system which has supported a highly developed religion and culture.'[21] He retreated to a Zen monastery and sat down to reflect. Here, he formulated the principles of a 'Buddhist economy', in which not greed but ethics, harmony and personal development were central.[22] Schumacher felt he had found an ideal society.

Unfortunately, the people of Burma were plagued by ethnic violence and their life expectancy hovered around 40 years.

When Ernst Friedrich Schumacher died in 1977, *The Times* declared ponderously: 'To very few people is it given to begin to change the direction of human thought.'[23] The newspaper considered Schumacher to belong to this minority. 'Small is beautiful' was the motto that made him into a big name. It became the title of his 1973 book, which at the end of the twentieth century was included in the *Time Literary Supplement's* list of the 100 most influential books since World War II.[24]

His plea for a smaller scale underpinned an intensifying call for setting limits to growth. We had to cut back: use less energy, buy less stuff, do away with our cars, have fewer children, close down industry.

For those in Europe and the US with an interest in alternative living and holistic thinking, Schumacher became an inspiration. They shared his desire for a lifestyle less dominated by consumerism. They showed a greater appreciation for the past, when nature was pure and morals unspoiled. They began to value tradition and longed for a connection with the local community. City dwellers began organising themselves into smaller groups – back to the village, back to nature. They grew food in their own organic gardens. They got their energy from ancient sources, such as wind and wood. As their own families grew smaller, they warned of population growth among those who weren't like them. And they searched for meaning, contrasting puny humankind, which was to be distrusted, with grand nature, which was to be admired.

If these beliefs had once been recognised as conservative, now they also became applicable to those who maintained their beliefs were progressive.

In *Dance with the Devil*, the far-right Günther Schwab, inspired by nationalism and conservatism, bemoaned the spread of material prosperity. He lets the 'Boss' Devil sigh: 'People no longer ask whether a thing is good or bad, but whether it is modern or out of date.'[25]

In *Small Is Beautiful*, E.F. Schumacher, inspired by socialism and Buddhism, describes 'the hollowness and fundamental unsatisfactoriness of a life devoted primarily to the pursuit of material ends, to the neglect of the spiritual'.[26]

Spot the difference.

Schumacher's vision resonated with others. In the early 1970s, *The Limits to Growth* became the title of an influential report by the Club of Rome, a think tank of mostly captains of industry. Young, ambitious critics of modern society emerged on the scene. During a discussion on nuclear power, Paul R. Ehrlich, a butterfly expert who wrote a bestseller on population growth in the human species, scolded: 'Giving society cheap, abundant energy would be the equivalent of giving an idiot child a machine gun.'[27] Criticism of mainstream thinking on energy prompted Amory Lovins, later the intellectual architect of Germany's *Energiewende* (energy transition), to propose an alternative: 'using as little energy' as possible.[28] Lovins was the first employee at Friends of the Earth, founded in 1969 with money from oil magnate Robert Anderson, former managing director of ARCO, now part of BP.[29]

Nuclear power, developed at a time of discomfort with technology and contempt for science, became the embodiment of an unnecessarily complicated technology – one surrounded by danger and secrecy, and enclosed by barbed-wire fences and warning signs. Was *this* what progress looked like?!

In a society seeking simplicity, there was little appreciation for boiling water by setting in motion a chain reaction of unstable atoms of enriched uranium. In people's minds, a coal-fired power plant was just a scaled-up furnace, a wind turbine was simply a modern version of the old windmill in the barnyard, and a beaver could build a hydroelectric plant if it was somewhat stronger. But... a nuclear plant? Were genius inventors like James Watt or Thomas Edison to travel to the future in a time machine, even they would have no clue what was going on in a nuclear reactor.

Around that time in the Netherlands, most people didn't have a clue either. This included Jannie Möller, who became known as the national 'mother of the anti-nuclear movement'.[30] In the early 1970s, she spoke to everyone from journalists to mayors. 'My role was to stir up trouble', she said in an interview in 2008. 'That was a request but I didn't even do it on purpose.'[31]

In 1970, on a visit to her sister's biodynamic farm in Germany, Möller read about nuclear energy from a brochure she found there. At the kitchen table, she learned that nuclear plants could explode and that radioactive waste was being dumped in nature. Leafing through the brochure, she became 'horrified' and convinced of her duty to inform the people of the Netherlands about this.[32] At the time she was visiting her sister, a nuclear plant was being built near the village of Borssele in the Dutch province of Zeeland, along the Scheldt River.

Möller looked into the author of the brochure and found the name Günther Schwab. She discovered his books and started reading. The next year, she was to meet the writer at a conference held by his WSL organisation. At the time of their meeting, Schwab was an honorary member of the German Cultural Work of the European Spirit, a far-right organisation for the protection of Germany's cultural heritage, and scientific advisor to the Society for Biological Anthropology, Eugenics and Behavioural Research, based on the National Socialist ideology of racial purification.

Back in Amsterdam, full of inspiration from her encounters with Schwab, Möller set up a working group, which included the figurehead of the anarchist counterculture; a WSL member whose Dutch translation of the Club of Rome report was selling exceptionally well; and a long-haired, newly graduated physicist named Wim Turkenburg, who over the following decades would transform into a prominent expert whose views on nuclear energy were seen as independent, even neutral. As a media personality in 2012, Turkenburg recounted how he became involved in the working group:

> I heard some people were looking for an author to write a report in the run-up to an environmental conference in Stockholm in 1972 … It had to be about oil. Working as a researcher on surfaces of solids, I knew nothing about oil, but I said yes, just out of interest. Then these people said: no, the report had better be about nuclear energy. All right then. I had no idea about nuclear energy either, but started reading magazines like Science and noticed how openly problems were reported. Problems with safety. With nuclear waste. Radiation.[33]

That year, 1972, the Dutch government announced a plan to meet some of the rising demand for electricity by constructing nuclear plants. Sitting together on a train, Möller and Turkenburg conceived the idea for an alternative memorandum. In a 70-page paper, they argued for reflection. They wrote that too much energy was being wasted, that the focus should be on energy conservation, and that a certain 'energy lobby' was blinded by the promises of nuclear fission. However, nuclear power, the authors continued, still raised many questions and involved many dangers.

Fortunately, there were alternatives, they wrote. But of all the 'natural energy sources', only solar panels 'can be expected to make a substantial contribution', and only 'in the long term'. Therefore, the authors concluded, a substantial expansion of 'fossil-fuelled central capacity' was better – *i.e.* coal and gas – because that would 'greatly benefit national well-being'.[34]

This was the origin story of the anti-nuclear movement in the Netherlands.

As soon as nuclear energy took off in the 1960s, warnings were issued, mainly from members of the environmental movement and the coal industry. Representing both groups, E.F. Schumacher labelled nuclear fission as 'an incredible, incomparable and unique hazard for human life'.[35] In *Small Is Beautiful*, he described nuclear plants as 'satanic mills', as if Günther Schwab himself reverberated in the pages of his book.[36] In another chapter, Schumacher (who in turn surfaced in Schwab's bibliography) lamented on page after page that too many British coal mines were closed in the 1960s, 'as if coal were nothing but one of innumerable marketable commodities, to be produced as long as it was profitable to do so and to be scrapped as soon as production ceased to be profitable'.[37]

Safety concerns were also heard from some within the nuclear industry. George Weil had worked on reactor designs and wrote an exposé in 1971. Weil, who removed the control rods one by one in the world's first nuclear reactor at the Chicago squash court in 1942, now deemed nuclear power to be dangerous and inefficient. Nuclear plants, he wrote, 'offer too little in exchange for too many risks'.[38]

Concerned insiders reported incidents or near misses that could have had major environmental consequences but were hushed up by the authorities. They were not so sure exposure to radiation was without harm and they criticised the nuclear industry's close relationship with the military.

Their candid reservations contrasted with the cocky attitude of the industry, which continued to boast of endless possibilities. For instance, the US Air Force squandered billions of dollars in fruitless attempts to use nuclear power in aircraft. There seemed to be no end to their future plans. Small atomic bombs must be detonated to dredge canals and a nuclear missile must be fired at the moon, just to see the effect.

Scepticism of nuclear miracles reached Hollywood. On 16 March 1979, *The China Syndrome* premiered in cinemas. Jane Fonda plays a television reporter, Michael Douglas a cameraman. They're filming inside a nuclear plant when an incident occurs. Alarms go off. The floor is shaking. Panic in the control room. For a while, it looks like the place could explode. In the end, it fizzles out.

Then attempts to dismiss the incident as trivial begin. The energy company obstructs an investigation into the events. Superiors at the TV station refuse to broadcast the secretly recorded footage. An independent expert looks at the film and concludes that everyone for miles around is lucky to be alive. When a plant employee notices a leaking pump and discovers that a safety report has been falsified, he is ignored.

Film critics raved. 'A terrific thriller that incidentally raises the most unsettling questions about how safe nuclear power plants really are', wrote one renowned reviewer.[39] The *New York Times* stated the film was 'as topical as this morning's weather report, as full of threat of hellfire as an old-fashioned

Sunday sermon and as bright and shiny and new-looking as the fanciest science fiction film'.[40] Critics were careful to mention that the fictional events were based on real mishaps, such as the stuck needle on a meter that caused technicians to misjudge the water level.

The nuclear industry bit back. The film was said to be full of fantasies, accusations and errors. An executive declared: 'The systems are designed and built in such a way that a reactor will operate safely even if there is a significant equipment failure or human error.'[41]

However, 12 days after *The China Syndrome* was released, alarms rang out in the newest reactor at Three Mile Island, a nuclear plant near Harrisburg, Pennsylvania. A minor malfunction in the cooling system had caused the reactor to shut down automatically. In the control room, hundreds of lights started flashing but none revealed that a relief valve had failed to close, causing coolant to drain and the reactor core to heat. Because the operators could not see on their instruments that the valve had remained open, their decisions resulted in a partial meltdown. Radiation escaped.

The next day, authorities insisted there was no cause for concern. Local residents listened and thought: *Yeah, right.* Some 140,000 chose to evacuate their homes for a while.

News about Three Mile Island reached Europe at a time when Germany was writing the next chapter in its protest against nuclear power. Resentment at the decision to store radioactive waste in a salt mine in Gorleben caused opponents to organise a multi-day march 160km (100 miles) long. It started with a few hundred hikers and cyclists. When the tour concluded in Hanover, newspapers were brimming with stories on what remains the most serious nuclear plant accident in the United States, leading to a turnout of at least 50,000 protesters in the German city.

Opponents of nuclear power were winning. With 50.5 per cent of the vote in a referendum, Austrians prevented the commissioning of a fully completed nuclear plant at

Zwentendorf an der Donau. The Swedes spoke out in their referendum against the construction of new nuclear plants in general. The Spanish declared a moratorium. One thing was clear: people did not want this. *Nuclear power? No, thanks!*

The industry crumbled. Construction plans for nuclear plants ended up in the shredder. After the Three Mile Island accident, pending orders for reactors came to a halt. Several nuclear plants under construction were converted into coal-fired facilities, without too much resistance from locals or environmentalists. A few nuclear plants that had already been built were mothballed before opening, such as the one on Long Island near New York City. After years left empty, the finished nuclear plant in Kalkar, Germany, was being turned into a theme park complete with Ferris wheel and ball pit.

Due to their success, protesters developed a taste for more. Green parties were forming in Europe on a base of anti-nuclear activists. By the 1980s, they were making a breakthrough, with Die Grünen in Germany being a pioneer. These parties developed a comprehensive programme around nature protection, global awareness and a cautious approach to technology. They broke with the style of traditional political parties, refusing to brand themselves as left or right, and found their electoral base mainly in progressive university cities. Everywhere, Green parties became the natural home for people who really don't like nuclear power.

Three years after his terrorist attack on a French nuclear plant, Chaïm Nissim entered politics as a representative for Grüne Schweiz, Switzerland's newly formed Green Party. By the time Nissim gave up his political work in 2001 and prepared for his confession in *L'amour et le monstre*, his colleagues in Germany and Belgium were part of a coalition government. They seized the opportunity for a firm decision: all nuclear plants must close. The Belgian government went one step further and enshrined in its constitution that there should never be another nuclear plant.

By the turn of the twenty-first century, the dissenting voice had become the dominant one.

Perhaps the very first developers of nuclear power anticipated the opposition their work would meet. The words they used for their discoveries had connotations of life and fertility, possibly in an attempt to evoke positive associations.

For instance, the term 'fission' originally comes from the biological process in which a cell splits in two, initiating new life. A reactor that produces more fissile material than it consumes is called a 'breeder reactor'. Witnesses to an atomic test described the moment as if they saw 'the Birth of the World' or 'the moment of Creation'.[42] An attack that could trigger war was once described as 'a big bang, like the start of the universe'.[43]

According to anthropologist Hugh Gusterson, these life-affirming metaphors served a purpose. They made the work of nuclear physicists a symbol of 'not despair, destruction and death but hope, renewal and life'.[44]

But when life and death compete, it's obvious which primal force holds more clout. Nothing could stand up to the opponents' death cult. During protests, people dressed up as skeletons, or as the Grim Reaper with his scythe. They compared nuclear plants to extermination camps like Auschwitz, while they placed signs on the fences bearing the words 'Death Factory'. They wore gas masks, carried coffins and held up pictures of the deceased and wounded from Hiroshima. They organised die-ins, where protesters lay on the ground as if dead. On posters against nuclear plants and nuclear waste, skulls often recurred, even deformed children without eyeballs.

In the United States, physicians gave lectures in schools, churches and community halls with the most gruesome descriptions of what would happen after a nuclear attack. Concrete, steel and glass would fly around. People would be decapitated, their lungs torn, their bones sticking out. Those who weren't vapourised immediately might be mortally wounded, with burns that needed to be treated for months, which would be extremely painful, but such treatment would likely be unsuccessful, not least because there would be too few doctors.

Most notably, a 1982 documentary short film that reproduced a lecture by Helen Caldicott, a paediatric physician from Australia, on the dangers posed by nuclear weapons, drew crowds and even won an Oscar.[45] In between the bouts of sobbing in the auditorium, you could hear a pin drop.

Opponents of nuclear power cultivated mortal fear. This became a key part of their strategy and a major reason for their success. The confrontation with death was disruptive. Who could trust the soothing reassurances of the authorities? Who really thought everything was under control?

The realisation that life and death are inextricably part of such a mighty form of energy was simply lacking on both sides of the debate. It may have been there among the alchemists of centuries past. They maintained that the true meaning of alchemy was not the literal conversion of lead into gold, but a purification of the soul. It's possible to renounce your sinful life, the alchemists argued, and start again as a virtuous person. To do so, you must take a journey through darkness and chaos. You have to abandon everything you own: all material things, all your loved ones, all your ideas and beliefs. If you then find inner peace, you will awaken spiritually.

The alchemists' mission was a personal one. Was it applicable to a society? Could it be that the dreadful atomic bomb had opened the door to a time of peace and progress? Was our knowledge of the basics of nuclear power a passage from death to life – from a world of disease and poverty, to a world of health and prosperity? Had that realisation perhaps finally dawned on Ernest Rutherford, the cautious founder of nuclear physics ('Don't call it transmutation'), not long before his own passage to death in 1937? The title of his last book: *The Newer Alchemy.*

Ah, the idea was absurd. Nor was there any time for such considerations. It was time for action!

Before long, opponents of nuclear energy would be presented with an even more powerful symbol of death. Shortly before the controversial nuclear plant near Brokdorf was finally due to open, more than 1,500km (over 900 miles) away, beyond the Polish border with the Soviet Union, something went spectacularly wrong in a nuclear plant seen as an example of the unrivalled technology of the socialist state.

PART TWO
OOPS!

CHAPTER FOUR

Doomed

What went wrong in the world's biggest nuclear disaster?

'It is a capital mistake to theorise before one has data. Insensibly one begins to twist facts to suit theories, instead of theories to suit facts.'

– Sherlock Holmes, in *A Scandal in Bohemia* (1891),
by Arthur Conan Doyle

Things looked bright for the residents of Pripyat. The town, built in the heart of nature near the old fishing village of Chernobyl, was a worker's paradise. Shops were well stocked. Thousands of young families eagerly awaited the festive opening of an amusement park.

Further along, a couple of tall, slim chimneys stood out, painted in red and white as if they were part of a candy factory. Over here, work was being done on the fifth and sixth reactors of the Vladimir Ilyich Lenin Nuclear Power Plant. This was a showpiece of Soviet might. When construction was complete, it would be home to the world's largest nuclear plant.

Viktor Bryukhanov observed it with delight. Back in 1970, he arrived with his wife and a job: along the river in northern Ukraine he would build an *atomgrad*, an industrial town for the workers in the nuclear plant. From the start, this was the Party's prestige project and he, Bryukhanov, at just 34 years old, was to help shape the socialist dream of electrification.

What Bryukhanov lacked in nuclear experience, he made up for with expertise gained in a coal-fired power plant. Surely it couldn't be that different? On top of an apartment building on the main square in Pripyat, he commissioned the painting of a propaganda slogan: 'Let the atom be a worker, not a soldier!'

Now aged 50, Bryukhanov was more proud of his work than ever before. It was April 1986, and Labour Day celebrations were coming up. Rumours abounded: for outstanding performance, the staff were expected to receive a bonus on 1 May. Bryukhanov himself might be pinned with the Hero of Socialist Labour star, the highest state decoration, possibly followed by promotion to Moscow, where, a year earlier, Mikhail Gorbachev took office as General Secretary of the Communist Party. Only a month ago, Gorbachev had spoken at a Party Congress about the need for *glasnost*, openness. A new era was dawning – a break with bureaucratic secrecy.

But it wasn't to be.

On the night of Friday 25 April, a safety test of reactor 4 failed spectacularly. The reactor exploded with a deafening roar. The 3m-thick (10ft) concrete roof was ripped open. The entire building shook. Debris fell from the ceiling, and the concrete walls buckled and bent. Pipelines jumped. Lights went out. Some thought: *the Americans have come.*

Chunks of hot uranium and graphite were scattered everywhere. Where a moment ago there was a reactor, now there was a boiling radioactive mush, open and exposed. The firefighters who rushed in couldn't put out the fires. Some became dizzy and vomited. Others suffered from dry throats and severe headaches. In the next hours, ambulances raced back and forth to the local hospital. There, some of the patients had swollen bright red faces. Instantly, the doctor on duty recognised the symptoms: this was acute radiation sickness.

A day later, more than 200 patients were flown to a hospital in Moscow. They were mostly technicians from the power plant and firefighters, but also guards who had remained at their posts, and construction workers who had been standing at the bus stop a bit further down the road, oblivious to the trail of radioactivity moving past them.

Information was missing. In Pripyat, it was decided, no one was allowed to leave town without permission. Telephone lines did not work. Radios installed in the walls of all homes to spread state propaganda made no sound. National television news did not report the events at the nuclear plant. Talk began to circulate.

Only on Sunday afternoon was the silence broken, via local radio. 'Attention, dear comrades!' There was an accident at the nuclear plant, it was said, and 'an unfavourable radiation situation is developing'. Thankfully, 'the necessary measures are being taken'. However, in order to stay safe, residents needed to 'temporarily evacuate'.[1]

That same day, an endless parade of buses took nearly 50,000 people to surrounding villages. When a traffic controller went to bed exhausted, she suffered a splitting headache and a sore throat. Her feet and ankles itched. As she had been arranging the evacuation, radioactive dust had blown around her bare legs.[2] Wind carried the dust particles along, further and further...

The next morning, in the early hours of a rainy Monday, the alarm sounded at the radiation monitoring point of a Swedish nuclear plant in Forsmark, north of Stockholm. A worker, on his way from the cafeteria to the changing room, noticed that the warning device was beeping at anyone who walked in. Could there be a leak in one of the reactors?

They found nothing.

Reports of radioactivity were now trickling in from Sweden, Finland and Denmark. Air samples were taken. In the laboratory, these samples were found to contain particles of graphite, a crystalline form of carbon not present in Scandinavian nuclear plants. The wind was coming from the south-east, from the direction of the Soviet Union.

Meanwhile, at a top-level meeting in Moscow, experts told Gorbachev that the Chernobyl accident would become apparent soon enough in Europe. Might this perhaps be an opportunity to fulfil his promise of openness? Some nodded, others frowned. Objections were raised. There might be panic, and nobody wanted panic. Nobody wanted to see a loss of prestige either. So when Sweden was on the phone that afternoon inquiring about a possible nuclear accident, the answer was that, no, *njet*, they knew nothing about it.

Only that evening did Radio Moscow confirm there had indeed been an accident. Details were sparse.

And all the while, agonisingly slowly, that invisible, mysterious cloud of radiation drifted across the European continent, through Mongolia towards Japan, and on to the US West Coast. Where rain fell, it was radioactive.

What happened at Chernobyl was new to everyone, and the Soviets weren't exactly spilling much information. Official messages from governments throughout Europe were contradictory. People were ordered to throw away milk and spinach, and children advised not to play in the sandpit, but the situation was under control. Really, nothing to worry about. Oh, but don't forget to throw away the milk and spinach, will you?

Precisely because of all the secrecy, the accident at Chernobyl became big news in the free West. Without a reliable source of information, journalists had to make do with speculation. Two thousand dead, one newspaper wrote.[3] No, 15,000, the same newspaper reported a few days later.[4] The bodies were said to have been dumped in mass graves.

What on earth was going on?

Whatever had happened, this was not simply an unfortunate engineering mistake at an industrial site somewhere far away. No, what had happened at Chernobyl, or so we began to think, was inherent in the technology of nuclear fission. This could happen again, anywhere. This accident would be a permanent source of harm, particularly to those nearby, but also across borders. And the worst part was the thought that the damage would accumulate unnoticed in their bodies. An epidemic of cancer was inevitable, it was said. Tens, no, hundreds of thousands of people, no, as many as a million would die from the effects of radiation exposure. And future generations would suffer too, not just in the Soviet Union but throughout Europe, even the world!

There was so much uncertainty, but we knew one thing for sure: *this was by far the biggest industrial disaster ever.*

It cannot be overestimated how much of an influence the Chernobyl accident had on the perception of nuclear power among those who watched or read the news. Over time, stories emerged about a wave of deformed children born in the wide vicinity of Pripyat in the aftermath of the tragedy. One was Igor Pavlovets, who was born in 1987 with one arm and stunted legs. He was placed in an orphanage and told that his parents had died from radiation. After a British couple adopted him, Igor became the focus of a 1995 documentary that aired on UK television. There were many like him. A voice-over stated: 'The genetic legacy of Chernobyl is a million deformed children.'[5]

In Germany, the little-known childrens' writer Gudrun Pausewang had just finished a book – *Die letzten Kinder von Schewenborn*, or The Last Children of Schewenborn – depicting life after a nuclear attack. There was no happy ending. Convinced that nuclear energy posed an existential threat, Pausewang decided that any nuclear plant could experience a disaster like the one at Chernobyl. Reading the news, she wondered: 'What would a catastrophe like this look like in the middle of the Federal Republic? I have to warn against that.'[6]

In 1987, Pausewang published *Die Wolke* (*Fall-Out*), a haunting read about a 14-year-old girl trying to escape the cloud of radiation after an accident at a nearby nuclear plant, but she is contaminated and loses her hair. There's no happy ending here either. The book became required reading in German schools, and shaped an entire generation's thinking on nuclear power.

Pausewang's apocalyptic writings may have been an attempt to reckon with her own conscience. In her teenage years, she had belonged to various Nazi youth organisations, believing in Hitler's message until the war came to an end. In an essay looking back on why she wrote *Die Wolke*, Pausewang acknowledged, 'I don't want to be asked by my grandchildren and great-grandchildren, like the grandchildren and great-grandchildren asked their parents after the Nazi era: "And you? Why didn't you do anything about it?"'[7]

In 2019, a new generation became aware of what had happened at Chernobyl. In *Chernobyl*, a miniseries by the leading US television network HBO with a stellar cast, we saw actors falling down by the dozen. A heroic scientist, in search of the truth, explained that we should think of every atom of fissile uranium as a bullet, and the reactor explosion as an atomic bomb that goes off 'hour after hour' and that 'will burn and spread its poison until the entire continent is dead'.[8]

While Soviet authorities set about upgrading all nuclear reactors of the same, unique Soviet design to meet international safety standards, the accident triggered something else among government leaders in other countries. They lost confidence in their own nuclear plants.

From Denmark and Austria to the Philippines and New Zealand, governments closed perfectly functioning nuclear plants and scrapped construction plans for new ones. After Italy held a referendum on nuclear power in 1987, the government proceeded to close its plants. The Swedish prime minister was clearest when speaking about Chernobyl, saying, 'Nuclear power must be got rid of.'[9]

Around Chernobyl's destroyed reactor, radioactive debris was removed and a huge composite steel and concrete shelter was built. Meanwhile, the other reactors at the plant were started back up, providing power as usual. But the world had changed for ever. Two years after the disaster, Moscow halted the construction of three nuclear reactors elsewhere in the country. Public pressure – a novel phenomenon since the Bolsheviks seized power in 1917 – could no longer be ignored. Thanks to a glimmer of *glasnost*, the people of the Soviet

Union occasionally caught a glimpse of the aftermath in their newspapers. The death toll, they read, was a few dozen at most. They didn't believe any of it.

Soon enough, the Soviet people came to realise that they had been deceived and misled, not only about what had happened at Chernobyl, but also about previous accidents at other nuclear plants and plutonium production sites for nuclear weapons that were now being cautiously discussed. Their country, it began to dawn on them, wasn't such a superpower after all. They felt despair and anger when they thought of the horrors under Stalin's rule; of nepotism in the Communist Party; of the senseless war in Afghanistan; of the everyday oppression to which they had become accustomed.

A good five years after the accident, it was not just the reactor at Chernobyl that had exploded. The entire Soviet Union had collapsed.

In 2006, Mikhail Gorbachev looked back on that episode. He viewed not his political reforms but the events at Chernobyl as 'perhaps the main cause' of his country's downfall.[10] The world's biggest nuclear disaster became the driver of the biggest political upheaval in recent global history.

Dwight Eisenhower got it right in 1953, albeit not as he had imagined, when he told the United Nations about the peaceful uses of the atom. Nuclear power, previously deployed to end World War II, had now helped to end the Cold War.

What exactly went wrong at Chernobyl? For his masterful reconstruction *Midnight in Chernobyl*, British journalist Adam Higginbotham ploughed through reports for more than a decade and spoke to endless experts, former employees and local residents. His conclusion: this is what complacency can do in a failing state when combined with disregard for nuclear power's real risks.

In their penchant for gigantomania, the Soviet nuclear specialists designed a reactor bigger and more powerful than any in the capitalist world. Their so-called RBMK reactors, using graphite as a moderator and water as a coolant, were cheap and easy to build. Also, you could – who would have guessed? – make plutonium for nuclear warheads. A complementary concrete dome, to prevent radioactive material from escaping the reactor building in case of a meltdown, was considered unnecessary.

'Nuclear power stations are like stars that shine all day long,' it was proclaimed.[11] 'We shall sow them all over the land. They are perfectly safe!'

Clearly, they were not.

A typical example of this confidence was a 10-page article on nuclear power that appeared in the February 1986 edition of *Soviet Life*, an English-language magazine intended to impress the outside world with all the splendour behind the Iron Curtain. Special attention was paid to Chernobyl. 'The plants have safe and reliable controls that are protected from any breakdown with three safety lines,' the Ukrainian energy minister said.[12] 'The odds of a meltdown are one in 10,000 years.'

Yet design flaws came to light from the outset. The RBMK reactor was unpredictable at shutdown. The emergency stop took as long as 18 seconds to have any effect. Inspections revealed flaws in the concrete. And whereas in Western designs the power output drops when the temperature rises in the reactor and the whole thing slows down, Soviet designs had it the other way round.

Problems were brushed aside. In fact, any criticism could mean dismissal. After all, hadn't the authorities said nuclear plants were perfectly safe?

Accidents did occur. In 1975, the first RBMK reactor, in Leningrad, had been in operation for less than a year when a concrete tank containing radioactive gases exploded. A month later the cooling circuit broke down, leaking radioactive water and killing three workers. Another few months later, radiation was released after damage to the fuel channel where water is converted to vapour.

Whenever an accident was investigated, it was always said to be due to a manufacturing error or staff inattention, never the design of the atomic scientists. If the information was shared at all, it was not with people working in the nuclear plant, and certainly not with local residents.

In the centrally planned economy, things went straight from the drawing board to the construction site. Completion dates for engineering projects were unrealistic, so everyone fiddled with and skirted around the rules. The roof of the turbine hall at Chernobyl was clad in highly flammable bitumen, but a fire-retardant alternative was not available in the Soviet Union, so it was tolerated with a blind eye. Because unemployment was viewed as failure in the workers' paradise, busloads of men and women arrived at the construction site with no idea of what they were supposed to be doing. Engineers and electricians had no specialist knowledge of nuclear plants.

Before commencing operation, the fourth reactor at the Chernobyl nuclear plant – the most modern, most advanced – was meant to undergo a key safety test. But as construction took longer than expected, the test was postponed to allow the job to finish in December 1983 as planned.

The following year, the test was suspended again.

And then again.

Meanwhile, in the control room, engineers were having great difficulty simply powering up and shutting down the reactors. The valves and flow meters proved unreliable. Adjustments were made by intuition. The control panel, with hundreds of switches, buttons, dials, meters and lights, soon showed wear and tear.

When at last the reactor 4 safety test, designed to demonstrate that the system could withstand an electrical blackout, was scheduled for 25 April 1986, formal permission had not been requested. That day, the test was cancelled after all, because workers would have had to shut down the reactor, while factories in the area needed every kilowatt to meet their production quotas before Labour Day. The testing would have to be done at night.

The night shift, however, was not prepared for the test. Upon entering, Leonid Toptunov, senior reactor control engineer, saw for the first time a stack of papers with instructions for running the test. He had only two months' experience operating the reactor. He was just 25 years old.

During the test, when reducing power, the reactivity meter unexpectedly dropped further than intended, almost to zero. While unplanned, it wasn't a recipe for disaster. They should have simply called off the test, shut down the reactor, left things as they were for a day, and then started it up again.

But according to deputy chief engineer Anatoly Dyatlov, who was in charge of reactor 4, the test had to continue. Here's how Higginbotham describes Dyatlov:

> He was rarely in his office but prowled the corridors and gangways of the plant day and night, inspecting the equipment, checking for leaks and errant vibrations, and keeping tabs on his staff ... He had no tolerance for shirkers or those who didn't follow his orders to the letter ... He could be high-handed and peremptory, peppering his speech with curses ... muttering to himself about the inexperienced technicians he dismissed as chertov karas – fucking goldfish. He demanded that any fault he discovered be fixed immediately and carried a notebook in which he recorded the names of those who failed to meet his standards.[13]

Much to his chagrin, Dyatlov had been waiting all day to supervise a test that kept getting pushed back. Enough already! Dyatlov ordered the power to be increased. *Pull up those damned control rods!*

Toptunov protested. Doing so would make the reactor go out of control. That couldn't be right, could it? Dyatlov, worn out after a long working day, responded with a threat. Did he have to go and get someone else to do the job?

Toptunov was stuck: disobeying would mean the end of his budding career. By obeying, however, more than just his career would end. When he withdrew the control rods, disaster was inevitable.

After the explosion, things got even worse. According to the dosimeters, the radiation levels didn't appear to be too bad. In reality, those instruments could only measure low levels of radiation. The dials went out completely, and the readings were then passed on as if they were measurements. Nobody dared to question the equipment or worry their superiors.

Some did not even want to acknowledge that the reactor had been destroyed, leaving men risking their lives in a futile attempt to supply cooling water. Gorbachev was told over the phone that the reactor – the one that was completely in ruins – would soon be restarted. Director Bryukhanov sat by, stupefied.

In the next few days, during endless meetings, firm decisions were made, simply to keep up the illusion that the situation was under control. However, no one told residents in Pripyat to close windows and stay indoors. No one advised them to avoid local milk in the weeks ahead. Even the hundreds of thousands of workers who were summoned to clean up the area for a considerable length of time were poorly protected from the radiation. Often they didn't even wear masks or gloves. Dust got into their eyes and mouths, and stuck under the clothes they wore through the night.

Mikhail Gorbachev, the darling of the West who was praised for his communication skills, remained silent. *Glasnost*, but not yet. Only after 18 days did he break the silence. In a televised speech, Gorbachev complained of 'a veritable pile of lies, the most unscrupulous and spiteful', all designed to 'denigrate' the Soviet Union.[14]

The government commission's official investigation into the disaster stated that nothing was wrong with the reactor design, rather the staff had made mistakes and supervision had been inadequate. Viktor Bryukhanov and Anatoly Dyatlov became scapegoats.[15] They had ignored protocol and were sent to a penal colony for 10 years. The former endured his humiliations resignedly, the latter bitterly.

While they were imprisoned, further investigations into the accident took place. The International Atomic Energy Agency (IAEA) ruled that the reactor design was flawed. One author of a report by the state's independent nuclear safety board claimed the cause of the disaster to be a combination of 'scientific, technological, socioeconomic and human factors' that occurred exclusively in the Soviet Union.[16]

Leonid Toptunov, who had operated the reactor and died from acute radiation syndrome on 14 May 1986, was posthumously awarded the Ukrainian Order for Courage. On his deathbed, he said in a weakened voice: 'Mama, I did everything right. I did everything according to the regulations.'[17]

How many lives did the accident at Chernobyl take?

The attendant at the circulation pump was killed instantly, vaporised on the spot by the heat, or perhaps crushed by the debris. A few hours later, a colleague who had been trapped in the chaos and could barely move his lips died of burns in hospital. Soviet authorities kept this total of two as the official death toll for a long time.

But it didn't stop there. For an estimate of the true death toll, we should turn to the best available scientific literature. In 2006, an updated version of *Chernobyl's Legacy: Health, Environmental and Socio-Economic Impacts* was published, written by the Chernobyl Forum, a collection of eight UN agencies, including the United Nations Scientific Committee on the Effects of Atomic Radiation (UNSCEAR), the World Health Organization (WHO), the UN Environment Programme (UNEP) and the World Bank, and supplemented by the governments of Ukraine, Russia and Belarus. With the participation of around 100 acknowledged experts, this is the most authoritative source on the Chernobyl disaster.

The report mentions that another person died of a heart attack at the scene.[18] That makes three.

Some 600 workers and firefighters required first aid after the explosion. In 134 of these cases, doctors diagnosed acute radiation sickness: a collection of symptoms that emerge after exposure to a high dose of radiation over a short period of time. They were tired and nauseous, vomited and had diarrhoea. For a while they seemed to recover, but then things got ugly, in keeping with the syndrome. Their lips became covered with thick, black blisters. Their gums turned red. They developed painful sores. Their hair fell out. Their skin turned red and purple, then brown-black before peeling off. Infections developed. The walls of their intestines were eaten away, ultimately leaving their bodies as bloody diarrhoea.

After months of gruelling treatment, most were on the mend, although cataracts and ulcers remained common. However, before the end of the summer, 28 patients would succumb to the effects of radiation and burns.

We are at 31 fatalities.

By 2004, out of that group of 134 people with acute radiation sickness, another 19 had died, from a variety of causes. Potentially linked to radiation were five cases of cancer, but there were also unrelated deaths from heart attacks, tuberculosis and car accidents. We regret the fate of these 19 unfortunate people, but, following the lead of the UN expert group, we must leave them out of our tally. The number still stands at 31.

Cancer takes time to develop. Thyroid cancer is a well-known result when radioactive iodine accumulates in the thyroid gland, especially in children. The iodine can get there if you drink milk from cows that grazed on pastures contaminated with radioactive fallout. And indeed, in the first few years after the Chernobyl accident, researchers noted an increase in thyroid cancer. By 2002, some 4,000 cases of the disease had been detected in people who were under the age of 18 at the time of the disaster and were among the more than eight million people living in the worst-affected areas in Ukraine, Belarus and Russia. Fortunately, thyroid cancer is highly treatable, and once it's been diagnosed, patients can usually be cured. Death from thyroid cancer occurs in exceptional cases, mostly when the disease goes undetected or is left untreated. Chernobyl Forum researchers made an estimated guess of 15 deaths among this group.

Now we are at 46 deaths.

But hey, it was still only 2006, the authors of the report cautioned. Fate may still strike. Some cancers may not appear until later this century. A precise prediction is difficult, because here we have to rely on complex mathematical models full of assumptions about what radiation does to the body. Long story short: in a group of over five million people including so-called 'liquidators' who were tasked with the recovery and clean-up, and local residents in 'contaminated' – the report has the word in inverted commas[19] – areas, we could expect around 4,000 additional cancer deaths up to the year 2065, mostly among those who worked on clean-up crews.

The report *Health Effects due to Radiation from the Chernobyl Accident*, published in 2011 by UNSCEAR, came to similar conclusions. No statistically significant increase in breast cancer. No measurable effect on fertility. No evidence of an increase in birth defects.

And that's it. In a population of many millions, no more than a few dozen identifiable, immediate deaths, plus in decades to come, a few thousand premature deaths from a disease that, for a variety of reasons, frequently occurs in old age, although it can, indeed, be linked to radiation exposure.

By comparison, in 1984, a gas leak at a Union Carbide insecticide plant in Bhopal, India, instantly killed some 4,000 people.

Countless heartbreaking stories, such as the one about Igor Pavlovets (who was reunited with his biological parents years after the British TV documentary) were investigated. They were sometimes fabricated, other times heavily contrived, but not one had a demonstrable or plausible link to radiation. Physical and mental abnormalities occurred before Chernobyl; they were no more common after. The increase in reports of birth defects in Belarus is attributed to better reporting, and it kept pace with an increase in other, non-contaminated areas.

However, rumours persist, including the story of the so-called Bridge of Death. Here, residents of Pripyat stood watching the fire at the nuclear plant, 3km (under 2 miles) away. They are all said to have died. In *Chernobyl*, HBO's miniseries, which the makers say is based heavily on a collection of personal anecdotes,[20] a young dad is standing behind a pushchair, watching the aftermath of the explosion from the railway bridge. Children look on in amazement at the snow swirling from the sky. In a later scene, we see the father with his baby in hospital, begging for help. Both have burns on their skin.

With regard to those who stood watching from the railway bridge, the *Chernobyl* end credits state 'it has been reported that none survived'. That wording is correct; 'it has been reported', indeed, but that doesn't make it true. In reality, this claim has been thoroughly debunked. The story of the Bridge of Death is an urban legend.[21]

There are many such tales. There was the apparent helicopter that crashed while the pilot was trying to dump a load of sand or boron to put out the fire in the reactor. And indeed, in *Chernobyl* we see a helicopter crashing, seemingly swallowed by the extreme radiation above the smouldering reactor. In reality, a helicopter crashed only months later, when the fire was long gone, after one of its rotor blades collided with a chain dangling from a construction crane.[22]

'They seem like small things,' Adam Higginbotham, the author of *Midnight in Chernobyl*, said in an interview, 'but there's this accretion of all these small things that are constantly repeated, that creates this mythological version of the Chernobyl accident.'[23] According to Higginbotham, the fabrications feed into our worst expectations of what a nuclear accident might look like. They're stories that are, he says, 'conveniently horrifying'.

But a pressing question arises. How is it possible that Chernobyl did not claim *more* lives?

Is it perhaps impossible to link health problems, illnesses and causes of death directly to radiation? After all, one cannot tell from a cancer cell why and how it was formed. Even complaints of the heart or the liver can have all sorts of causes. Maybe some of these deaths can be traced to Chernobyl?

Not likely, experts say. In the years since, there has been no marked increase in any disease in the wider area that can with any justification be traced to the disaster. Still, it doesn't stop people from attributing a range of ailments to radiation exposure, such as tuberculosis or hepatitis, even though these are infectious diseases caused by viruses.

Nevertheless, while plenty of anecdotes may not add up to scientific evidence, they do seem to indicate *something*. Anatoly Grishchenko, one of the helicopter pilots who spent days dumping sand over the reactor, died of leukaemia in 1990. Was his death caused by the disaster? That's difficult to determine. According to all the best available knowledge – based on the long-running study of survivors of the atomic bombs dropped on Hiroshima and Nagasaki – leukaemia typically occurs between 10 and 15 years after exposure; before and after that, the likelihood of leukaemia is dramatically lower. Perhaps it formed in Grishchenko's case much faster than usual? Maybe he is indeed among the few thousand extra cancer cases. Maybe not.

What about Alexander Yuvchenko? He was a mechanical engineer in Chernobyl who received blood transfusions and skin transplants in hospital. After a year, he was allowed to go home. He died in 2008 from leukaemia. Was his cancer due to the disaster? It could have been, but again it's tricky, as we're now more than 20 years on. It could just as easily have been caused by something else: hereditary predisposition, exposure to chemicals, treatment with certain drugs. We will never know the true cause of his disease.

So doubt remains.

Another theory on why the death toll is so low is that doctors have been reluctant to attribute their patients' symptoms to the accident. A diktat from Moscow stipulated that the word 'radioactivity' was forbidden if patients were not seen to be overtly suffering from radiation sickness and burns. This led to the popularisation of a new diagnosis: vegetative-vascular dystonia, an amalgam of a whole range of vague symptoms, such as palpitations, headaches, depression, fatigue, irritability, nausea, dizziness, sweating, wheezing, coughing and frequent urination. And so while the diktat suggests that damage to health has been deliberately glossed over, according to the medical studies these complaints cannot be due to radiation.

Could it be that it's difficult to uncover reliable information from a country that had become used to falsifying and obfuscating documents? It's hard, indeed, except that there were also droves of international investigators on the ground. They were put to work with vast amounts of research funding from Western governments and European institutions. If anything, it would be tempting for the researchers to report on an increase in all kinds of diseases, because it would add importance to their work, and more research money and commissions would come their way.

But no matter how they searched, they found nothing.

Unless there was a co-ordinated cover-up… Did the authorities try to whitewash the truth? It wouldn't be the first time. In 1982, there was a partial meltdown at Chernobyl. Following maintenance, a cooling valve remained stuck. Once the reactor was turned back on, a tank overheated and ruptured. The accident was only noticed hours later when radioactive particles had already escaped through the ventilation system, carried along by the wind and precipitated by rain. A clean-up operation was required, but the accident was not made public.

A cover-up, then, isn't such a crazy idea. It's quite possible that the owners of the Three Mile Island plant in Harrisburg, Pennsylvania would have liked to cover up the accident at their plant back in 1979, just like in the film *The China Syndrome*. Such a course of action is not alien in any line of business, but especially appeared ingrained in the nuclear industry, with its tendency towards secrecy and mendacity.

Two accidents in the autumn of 1957, at a time when everyone was still enchanted with nuclear power, demonstrated this. First, an underground tank with plutonium waste exploded at a Russian secret military site deep in the Urals. Soldiers, bleeding and vomiting, were taken to hospital. A drizzle of thick, black snow fell on the surrounding villages. Yet the Soviet authorities stayed silent. Some 10,000 local residents had to guess why they were evacuated. For, how could anything have happened in a place that, officially, did not exist?

Less than two weeks later, fire broke out at the newly opened nuclear plant near Windscale, Cumbria, along the English coast. While the fire persisted, a typical newspaper article such as the one in *Shields Daily News* reported reassuringly that, according to state authorities, 'continuing measurements outside the site confirm there is no evidence of any increase in radioactivity which might have caused harm to the public'.[24]

At first the British government played down the events, stating that the radiation was completely harmless. Then it started to backpedal: farmers far and wide had to throw away their milk for weeks to come. Finally, it withheld the truth: Prime Minister Harold Macmillan ordered investigative reports into the fire to be confiscated or heavily censored before publication.

Covering it up – that seems to be the obvious response to any nuclear accident, anywhere. So was the truth about Chernobyl covered up?

Yes, believes Kate Brown, a professor of science, technology and society at the Massachusetts Institute of Technology, and a strong proponent of oral history. In her 2019 book *Manual for Survival*, she suggests that there were at least hundreds of thousands of deaths after Chernobyl, a claim that had long been buzzing around in certain circles. Soviet researchers, writes Brown, were forbidden to report on the 'stunning increases' in cancers, birth defects, child mortality and many other conditions.[25]

Brown also suggests that the United Nations has done everything in its power to suppress the truth about the scale of the disaster. The UN is determined to withhold the facts, Brown argues, because its main member states possess nuclear weapons, and if the truth about the Chernobyl radiation deaths were accepted, these countries would also have to acknowledge that radioactive fallout from atomic testing has led to what she has named the 'most suicidal era in human history'.[26]

Evidence of mass deaths and of attempts to suppress the truth is lacking; alternatively, it must be that UN researchers ignored studies of questionable quality. 'Somewhat selective,' is her assessment of their work in an interview.[27] 'They tossed out this work they didn't like.'

Through her theories, Kate Brown builds on a debating style that has become popular among opponents of nuclear power. A famous exponent is Helen Caldicott, the Australian physician who rose to fame in the 1970s through her passionate fight against nuclear weapons before gradually turning her sights on nuclear plants. Caldicott called the studies on Chernobyl 'a total cover-up' and 'the biggest medical conspiracy in the history of medicine'.[28]

There are parallels with other public debaters who make bold, even outrageous allegations to cast doubt on scientific conclusions. For example, there are the claims that vaccines are unsafe and that the pharmaceutical lobby dictates the outcomes of studies. Those who say such things rarely play a significant role in public debate.

And that's the odd thing: whereas deniers of vaccine safety are met with widespread disapproval and seen as conspiracy theorists, unfounded exaggerations of the consequences of Chernobyl are remarkably often endorsed. Brown's book received rave reviews in *Science* ('a must-read')[29], *Nature* ('a page-turner')[30] and *The Economist* ('a magisterial blend of historical research, investigative journalism and poetic reportage').[31]

The TV documentary about young Igor Pavlovets, symbolising 'a million deformed children', won an award at the Prix Europa, the biggest awards festival for European media producers.

Die Wolke became a bestseller and was awarded several literary prizes, while author Gudrun Pausewang received the German Youth Literature Award for her life's work and was awarded the Order of Merit of the Federal Republic of Germany.

HBO's *Chernobyl* went on to become the series with the highest audience ratings ever and was showered with Emmy and Golden Globe Awards.

The observation that the radiation released after the world's biggest nuclear disaster actually caused very little death and destruction raises uncomfortable questions. When looking for answers, some develop an unwavering suspicion that hundreds of experts associated with various United Nations agencies and national governments, plus a whole series of doctors, researchers and government officials from different countries, spent decades successfully concealing the cause of death of hundreds of thousands or even many millions of people.

Others raise an entirely different question, which may come across as equally crazy: could it be that radiation is not nearly as terrible as we think?

CHAPTER FIVE

A Strange Glow

How dangerous is radiation?

'*Mysteries are not necessarily miracles.*'

– Johann Wolfgang von Goethe, in *Goethe's Opinions
on the World, Mankind, Literature, Science, and Art*, 1853

If you're reading this book at home, radiation is hitting you
from the rocky materials in the masonry in the walls. If
you're outside on a bench, radiation is coming from the
cosmos above, as well as the earth below. When gardening
and digging in the ground, or when going on holiday in the
Alps, you increase your exposure to radiation. If you go by
plane, you receive even more. Even if you stay in the UK and
drive an hour from, say, Swindon to Bristol, or from central
to east London, the radiation level will double.[1] Want to avoid
radiation completely? Good luck. You will have to drive a
little further and leave this universe.

Much of this so-called ionising radiation can be traced
back to the Big Bang. Radiation is therefore not some
unnatural phenomenon we modern humans have introduced
into our living environment. On the contrary, it has made us
who we are.

Some of that radiation gets into our bodies. At the end of
each day, we have consumed a few more radioactive
substances, for example, when we eat meat or fish, or add
garlic to a pasta sauce.

In other words, we ourselves are radioactive.

However, a portion of the radiation we bathe in every day
is not natural. It's in the atmosphere due to the fallout from
atomic tests and the meltdowns of nuclear reactors, even now,
after decades.[2] A much larger piece of that artificial radiation

– and this is a part that is steadily increasing – reaches us with our consent, such as when we submit to certain kinds of medical imaging or radiotherapy in hospital.[3]

Yet there is something eerie about radiation, whether it comes from a nuclear reactor, the cosmos, garlic or a CT scan. Just the word 'radiation' makes us feel uncomfortable. We cannot see it, hear it, feel it or smell it, yet it can spread all over the world. *How can we ever be safe?*

Radiation is a special force of nature. It can make us sick, or cure us. It can kill us, or save our lives.

This ambiguity also struck the introverted professor of physics at the University of Würzburg who stumbled upon its discovery. On a Friday afternoon in November 1895, Wilhelm Conrad Röntgen conducted an experiment in the laboratory below his living room using, among other things, a glass tube, a piece of cardboard, a filament, a metal plate and an electric current. That day, however, the light he created under atmospheric pressure – as many of his contemporaries did – illuminated more than just the tube. A faint yellow-green glint appeared on a screen of barium salt that happened to be nearby. Yet the tube was wrapped entirely in black cardboard. How could that be?

Over the weekend, the professor continued to experiment. A mirror did not deflect the invisible rays. They passed right through cardboard, a thick book, a metal plate and a piece of wood. Only lead could stop them. Holding his hand between the tube and the screen, Röntgen saw a blurred outline of his bones. The image horrified him.

No law of physics could explain the mysterious behaviour of these powerful rays. When writing an article about his findings, Röntgen didn't quite know what they were or what to call them. He referred to them as 'X-rays'. Against his wishes, others started calling them 'Röntgen rays'.

For a long time previously, it hadn't looked as if Röntgen would make one of the most important scientific discoveries of his time and receive the very first Nobel Prize in Physics. At school, Wilhelm was not a high-flyer. The son of a

Dutch mother and a German textile merchant, he studied at the Technical School in Utrecht, where his performance in physics was assessed as very poor. After being expelled as punishment for allegedly drawing an unflattering cartoon of one of his teachers, he was refused admission to the university; his command of classical languages was insufficient.

Röntgen moved to Switzerland, where he obtained his doctorate and met the woman he later married. His wife, Anna Bertha, appears in all subsequent stories about Wilhelm Röntgen, as she lent her left hand for a photographic image using X-rays. When she saw the photograph of her bones, her wedding ring clearly visible, she became upset. According to legend, she exclaimed: 'I have seen my death!'

Röntgen's discovery didn't go unnoticed. Articles appeared in newspapers everywhere. Doctors in particular realised its practical usefulness. Now they could look inside their patients' bodies without cutting into them. Within weeks, Röntgen received a royal award. Within two months, there was the first medical application. A man in Canada had been shot with a revolver on Christmas Eve, somewhere in his leg. But: where was the bullet?

According to the prevailing custom of the time, doctors would wriggle their fingers into the wound, but they couldn't always find what they were looking for. This method was not without danger, as disinfection was not yet in vogue; only 15 years earlier, then-president of the United States, James Garfield, had lost his life after a doctor used his unwashed fingers to look for a bullet between muscles and tendons, causing a deadly infection.

But now everything was different. Now we had X-rays!

The leg of the stricken Canadian man was photographed, revealing what was inside. The photo was slightly underexposed, but the bullet was detected and removed.

Yet the principles behind the X-rays were still anyone's guess. The answer came, unexpectedly for the time, from a woman – the first to establish herself in modern science and who would go on to become the most famous female scientist in the world.

Growing up in Warsaw, Poland, Maria Salomea Skłodowska, like all girls, was forbidden to study. Her country was under the thumb of the Russian empire, which had annexed Poland. The natives were not even allowed to speak their own language. Maria's father, a widower since losing his wife to tuberculosis, allowed her to study secretly in the evenings after she'd completed her work as a nanny for affluent families during the day. In 1891, she was admitted to the University of Paris.

Paris! The brand new Eiffel Tower rose above everything else in the city. Underground, the Metro was coming. Streets and fountains were electrically lit. As the modern world took shape, Marie, as she was now called, lived on bread, butter and tea. Her room had no heating; in winter, she slept with all her clothes on, including a thick coat.

After graduating in physics, Marie went to work in a grubby laboratory she shared with Pierre Curie, an older physicist who built his own measuring equipment and still lived with his parents. In 1895, a year after they met, they got married. Their honeymoon was a bicycle ride.

A little later, when Marie Curie was looking for a topic for her dissertation, she learnt of Wilhelm Röntgen's X-rays. Everyone was talking about them. One of her lecturers, Henri Becquerel, had tested his entire collection of fluorescent minerals – a hobby of his – and discovered that uranium emits X-rays. Becquerel spoke of '*rayons uraniques*' (uranium rays). Curie changed the term to 'radioactivity' when she examined all kinds of minerals, metals, salts and oxides in the laboratory and found that thorium produces the same rays. X-rays are not reserved for uranium. Perhaps, she suspected, radioactivity is a property of atoms?

More research was needed. The university offered her an abandoned wooden shed with a leaking roof where, not long before, corpses had been cut open for autopsy. Here the Curies had a shipment of uranium ore delivered from the Bohemian town of Joachimsthal, where uranium was abundant in the silver mines (see Chapter 2). The stuff had been dumped in a pine forest. The owner had it brought over by horse-drawn cart, charging only for expenses. The bags were still full of pine needles.

Then the hard work began. Some days, Marie poured litres and litres of chemicals and gases into a cast-iron pot, stirring the boiling goo for hours with a huge iron rod. On other days, she sat down to the delicate work of solidifying the liquid substance. In this way, she purified two radioactive elements in 1898. One she called polonium, after her native country; the other, radium, derived from the Latin *radius*, for ray.

Due to their love of science, the Curies did not patent their discoveries. They were proud of their work. Pierre put bowls of radium in the courtyard near the lab; the extraordinary blue glow impressed visitors. Marie often carried a vial of radium in the breast pocket of her lab coat, just to take a look at it.

Their work caused a shockwave in the scientific community, just like the earlier discovery of X-rays had. Radiation, we now know, is not a result of interaction between two elements, but rather the result of the constant and gradual release of unprecedented amounts of energy stored in atomic nuclei in the interior of certain elements.

How could that be? The Curies did not know, despite sharing a Nobel Prize with Henri Becquerel for their discovery of radioactivity in 1903. The explanation for the peculiar phenomenon came two years later thanks to a 26-year-old man who neither taught at a university nor worked in a laboratory, but was an assistant in a patent office in Bern, Switzerland, in the Electromagnetic Apparatus Department. His name? Albert Einstein.

In 1905, Einstein published no less than four seminal papers that turned the field of physics on its head. In one, he laid the foundations for the most famous equation in history, penned in the *Jahrbuch der Radioaktivität und Elektronik*: $E=mc^2$. It offered an explanation for the immense energy that can be released when an atom is cracked open, as later seen in nuclear fission. The energy (E) coming from atoms as radiation is the mass (m) it has lost, multiplied by the square of *celeritas* (c), the speed of light.[4]

The worldview shifted. Reality, Einstein taught, consists not so much of matter as of time and space. We can all take comfort in knowing that even Einstein himself struggled to understand what all this meant.

In the case of Hans Castorp, the main character in Thomas Mann's 1924 masterpiece *The Magic Mountain*, his experience of radiation began when he had an X-ray of his chest taken at a sanatorium:

> *They heard a switch go on. A motor started up, and sang furiously higher and higher, until another switch controlled and steadied it. The floor shook with an even vibration. The little red light, at right angles to the ceiling, looked threateningly across at them. Somewhere lightning flashed. And with a milky gleam a window of light emerged from the darkness: it was the square hanging screen, before which Hofrat Behrens bestrode his stool, his legs sprawled apart with his fists supported on them, his blunt nose close to the pane, which gave him a view of a man's interior organism.*[5]

It must also have been like this in 1912, when Mann's wife was admitted to a sanatorium in the Alps with vague symptoms. And so, straight after the invention by Wilhelm Röntgen – who sought no fame, never wanted to profit financially from his discovery and would die penniless – hospitals had their own X-ray machines built. Some of these contraptions looked like enormous cabinets you could stand in. They creaked and they popped. Sparks flew around. In that atmosphere, patients had to remain motionless, sometimes for an hour as exposure time was not yet measured in milliseconds.

But the result was awe-inspiring: a print of everything that was or was not well in the body. Doctors could locate kidney stones, bone fractures and tumours. Dentists could detect dormant cavities between teeth. Spiritualists proclaimed they could photograph the soul. There was a rumour that special glasses allowed one to see through clothes; one manufacturer advertised women's underwear that protected against this.

The definitive breakthrough for X-ray photography came during World War I. Marie Curie herself used the technique on the battlefront by equipping cars with X-ray instruments, which became known as *petites Curies*. Then she trained individuals, including her daughter Irène, to take X-rays of countless soldiers to detect bullets and shrapnel in their bodies and help with their treatment. Lise Meitner, the Jewish scientist who pioneered uranium fission with Otto Hahn in the late 1930s (see Chapter 1), did the same work for Hitler's allies in the Austrian army.

Visiting his sick wife in the sanatorium, Thomas Mann saw the thin line between sickness and health, between life and death. What did the nervous Hans Castorp see in *The Magic Mountain*?

> *And Hans Castorp saw, precisely what he must have expected, but what it is hardly permitted man to see, and what he had never thought it would be vouchsafed him to see: he looked into his own grave. The process of decay was forestalled by the powers of the light-ray, the flesh in which he walked disintegrated, annihilated, dissolved in vacant mist.*[6]

Mann gave a literary twist to the panic that once overtook Frau Röntgen. But where they detected death, others observed life. Even then, radiation was used for more than just taking pictures. A drop of radium was inserted by injection or capsule into patients suffering from everything from tuberculosis to cancer. Even hairy birthmarks were radiated away. Pharmacists sold dozens of medicines with radioactive ingredients. They promised protection against bacteria and viruses, and the promotion of blood circulation and sexual energy. In the 1920s and 1930s, it was widely believed that radiation enhanced your health.

Spas, which by now were known to obtain the heat in their natural baths from radioactivity in the soil, noticed a boost in visitor numbers. Joachimsthal, the source of uranium used in the Curies' scientific work, attracted thousands of guests from the bourgeoisie in Vienna and Prague every year. Many stayed at the Hotel Radium Palace, where they drank the locally brewed Radium beer in the lobby. One of the guests was Robert Oppenheimer, who, as the 'father' of the atomic bomb, would later say that his interest in science began when his uncle gave him a collection of colourful stones from the Joachimsthal uranium mines.

A craze ensued. Radioactive ingredients appeared in soaps, salves, bath salts, face cream and hair growth remedies. You could snack on radioactive chocolate, and then brush your teeth with toothpaste that contained a little thorium, for a sparkling smile. When Germany began research towards an atomic bomb in 1939, the War Ministry searched for fissile material in a toothpaste manufacturer's landfill.

Radium became synonymous with modernity, vitality and quality. So popular was radiation in the public imagination that many products claimed to be radioactive even if they weren't, for example, 'Radium' condoms in America, which were 'radioactive' in name only. Meanwhile, doctors tried radium on patients with heart ailments, infections, high blood pressure, epilepsy, headaches, diabetes, arthritis, rheumatism – the entire medical encyclopaedia, really.

In all this optimism, an elixir of life was not inconceivable. An American entrepreneur produced bottles of distilled water with radium, called Radithor, which became a resounding success. 'Certified Radioactive Water', the label advertised. According to the salesman – a former Harvard University dropout posing as a doctor – it worked well for the treatment of pain, asthma, diabetes, constipation, impotence... well, against anything.

It also worked well if you wanted your bone tissue to wither away. An avid user, who drank Radithor throughout the day for years, lost his teeth and jaw before holes appeared in his skull. After a long and agonising struggle, he died of cancer in 1932.[7]

By then, the risks of radiation were well known. Victims of overexposure to radiation felt weak, developed cataracts, lost their hair or became temporarily infertile. The damage manifested itself mostly in doctors and their assistants, who experimented lavishly without proper protection. Some tested X-ray equipment by holding their arm in front of the screen. Many of their patients benefited from the diagnostic radiation, but they themselves suffered badly. High-dose radiation damages DNA. When proteins cannot repair that damage, cancer can occur.

The risks also became apparent when luminescent paint was applied to watch dials worn by US soldiers in the European trenches, so that they could see the glowing numerals and hands in the dark. Millions more were sold after World War I. The dials were painted in workshops by young women, who began to develop all kinds of ailments. Their teeth loosened and sores appeared on their gums. One felt pain just from touching her face. A dentist pulled out a suspicious molar. A piece of her jaw came with it.

What was going on?

The key ingredient in the paint was radium, to which chemicals such as zinc sulphide had been added to make it glow. To sharpen the tip of the brush for the delicate drawing on the dial, the women would routinely put it between their lips. That was how they had been taught. It was no more than a thousandth of a gram of paint each time, but still a teaspoon a week, a coffee cup a year. Sometimes the women painted their cheeks or teeth with it, just for fun. Thus, more and more radium stuck to their bones. By 1923, the first deaths were recorded in a series of workers who would succumb to a new occupational disease: radium jaw.[8]

Employers denied any responsibility. One spread the rumour that the women had contracted syphilis before falsifying a damning report, so that it appeared as if the entire workforce was in perfect health and the workplace was spotless. In reality, everything and everyone in the workshop gave off light: from the chairs the women sat on to their hair and corsets.

The workers began a lawsuit. Their bosses, it turned out, knew about the dangers and had deliberately concealed them. The legal action was widely reported in the press. Eventually, in 1928, the women were awarded damages. The industry would also have to introduce safety procedures and enforce them.

It was a historic victory. The case led to legislation requiring employers, including in other industries, to take safety measures and give workers the right to claim damages for negligence. Also, at the end of the court case, radiation protection became a scientific discipline with the creation of what would become the International Commission on Radiation Protection (ICRP), an independent organisation that would protect humans and the environment from the harmful effects of radiation by setting standards and making recommendations.

This short history from Röntgen to the 'radium girls' illustrates how radiation has always evoked mixed feelings. Both positive and negative emotions are deeply rooted, as Spencer Weart convincingly describes in his brilliant cultural history of nuclear power, *The Rise of Nuclear Fear*. There is a subconscious hope that radiation can work miracles. In folk legends, rays were linked to procreation and healing powers. In religious symbolism, the body of a divine or holy person was surrounded by a halo of rays.

At the same time, there was the age-old dread that radiation posed danger. It used to be feared that evil spirits could spread their bad influences through rays from their 'evil eye'. Witches could cast a spell on anyone through invisible forces.

Sometimes the positive and negative emotions surrounding radiation followed each other in quick succession. Thomas Edison, in his time the most famous inventor in the world, with the most patents to his name, saw the potential of X-rays. He had his team work on a fluoroscope, a device that allows you to look inside the body. Clarence Dally, a loyal laboratory assistant, provided his left hand for tests and demonstrations. When burns started to appear on his skin, he simply used his other hand. The wounds would not heal. At night, Dally slept with his hand in a tub of water to help with the pain. His hair fell out, including his eyebrows and eyelashes. His hands swelled. As the cancer advanced, he was forced to have both his arms amputated. The disease continued to spread throughout his body until Dally died in 1904.

During his assistant's illness, Edison abruptly stopped working on the fluoroscope; although it is still used in medical operations today. He wanted nothing more to do with it. When asked in an interview about Dally's situation, Edison said: 'Don't talk to me about X-rays, I am afraid of them!'[9]

The discovery of radioactivity raised all sorts of unconscious associations. A series of books, comics and films appeared in which radiation bestowed magical powers. This went both ways: unscrupulous monsters wreaked havoc with their deadly rays, but superheroes possessed superpowers thanks to radiation.

This ambivalence was reflected in the character of Superman. His X-ray vision allowed him to see through walls, but he was terrified of the radiation from kryptonite, which could incapacitate him.

Our own superheroes from Paris, the Curies, began to develop health problems due to their work with radioactive sources. Pierre sometimes applied radium to his arm so he could study the burns. At times, he suffered excruciating pain, though his death in 1906 came not from radiation but from crossing a rain–soaked cobbled street and slipping, after which the wheel of a horse cart crushed his head. Marie also regularly suffered burns on her fingers and hands. After her fieldwork during the war, her cataracts worsened and she became frail. She died of anaemia, perhaps caused by bone marrow failure.

Just before her death, Madame Curie was visited by her daughter Irène, who had also become a scientist. She had great news for her mother. With her husband, Frédéric Joliot, she had bombarded elements with ionising radiation, creating radioactive isotopes. Boron turned into nitrogen, magnesium into silicon, aluminium into phosphorus. It meant that Irène Curie and her husband were the first to have deliberately created radioactivity.

At long last, the alchemist's dream had come true.

Marie Curie died in a sanatorium in the Alps on 4 July 1934. That day marked a transition in the history of the atomic age: the period of scientific curiosity gave way to one of opportunism. As historian Richard Rhodes has noted, it was the same day that, in London, the Hungarian intellectual Leó Szilárd (see Chapter 1) filed a patent for the nuclear reactor.[10]

Since those early experiments with radioactivity, we have come to learn a lot about it. We know how radiation can be useful in healthcare. Targeted X-rays are used to diagnose and cure all kinds of diseases. With precise equipment, cancer cells are irradiated and killed, while healthy cells are spared. Dozens of radioactive isotopes are used both to diagnose and to treat diseases.

Medical isotopes to diagnose and treat tumours are made in a nuclear reactor. Their names appeal little to the imagination. Iridium-192 is used to irradiate cancer tumours. Lutetium-177 is for the treatment of metastatic prostate cancer. Molybdenum-99 is the raw material for technetium-99m, which controls blood flow in tissue. These and other medical isotopes benefit tens of thousands of patients every day.

Radiation is also used outside the healthcare sector. Older smoke detectors contain americium-241. The luminous green signs indicating escape routes in buildings contain tritium gas. Space probes and satellites in the solar system are powered by plutonium. Selenium-75 is used to detect cracks in materials or failed welds in buildings, bridges and aircraft.

Not only do we know more and more about how to put radioactivity to good use, but we also know more and more about when it is harmful. When radiation enters the body, damage can occur to cells and organs. But not all radiation is the same and not all radiation is equally dangerous.

In 1898, Ernest Rutherford discovered three main types of naturally occurring radiation by studying radioactive decay: he named these alpha, beta and gamma radiation. Alpha radiation immediately gives off much of its energy and cannot even penetrate a sheet of paper or a layer of dead skin cells. However, if alpha rays do penetrate the body, for example after ingestion or inhalation, they can damage living cells. This happened to the former KGB officer Alexander Litvinenko, a prominent critic of the Russian president Vladimir Putin, who was poisoned in 2006 when his tea was infused with a large dose of polonium-210.

Beta radiation contains much less energy than alpha radiation, and it barely gets further than a few metres when it travels through the air. It passes through paper, but not through an inch of aluminium.

Gamma radiation is an electromagnetic radiation like light, and is much more penetrating than alpha or beta. Gamma radiation cannot be completely blocked, but it can be very much attenuated. To reduce gamma radiation emitted from an operating nuclear reactor to acceptable levels, you need 30cm (12in) of lead, 2m (6.6ft) of concrete or 5m (16ft) of water. When it's used in a hospital, a syringe is surrounded by a protective sleeve containing 2mm (0.08in) of lead.

Radiation exposure and radioactivity can be measured in a remarkable number of units, such as the roentgen, the becquerel, the curie, the gray, the rem and the rad. When determining the biological impact of different types of radiation on living tissue in humans, the sievert is the most common unit. Named after the Swede Rolf Sievert, who worked at the intersection of physics and medicine, the sievert refers to the 'effective dose' of radiation. This expresses the risk a person faces after receiving a dose. The sievert is on the large side. Therefore, doses are usually reported in one thousandth of a sievert: the millisievert, or mSv.

Now for some figures: how much radiation leads to what consequences? Numbers are important. A famous theorem of Paracelsus, the sixteenth-century physician, alchemist and founder of toxicology, states that it is the dosage that determines whether something is toxic.[11]

When the radiation in our bodies exceeds 5,000 mSv, death follows in a matter of weeks or months in half the cases. A series of symptoms precede it: high fever, internal bleeding, infection, diarrhoea, loss of appetite, dehydration. In Hiroshima and Nagasaki, some of the victims received as much as 20,000 mSv in one blow. All of the cells in their bodies died off at breakneck speed. When that happens to nerve cells in the brain, all systems in the body break down. The victims are likely to have welcomed death. Indeed, Leslie Groves, the military director of the Manhattan Project, was wrong when he declared after the war: 'In fact, they say it is a very pleasant way to die.'[12]

A sudden dose of 1,000 mSv is likely to cause burns, but the symptoms of acute radiation sickness are no longer inevitable. The cells in the body are less likely to die. The body will recover because enough cells remain present in the marrow to grow and make new blood cells.

Around 100 mSv is the limit at which we reasonably assume some tissue damage will occur. That damage manifests itself in cell mutation, a potential harbinger of cancer. Because there are many causes of cell mutation, and thus many causes of cancer, it is impossible to determine at the individual level whether radiation is the culprit.

Up to 100 mSv, no evidence of persistent health damage has been determined.

As we saw earlier, we need not be near an exploded atomic bomb or nuclear reactor to be exposed to radiation. According to the UK Health Security Agency, residents of the United Kingdom receive an average of 2.7 mSv of ionising radiation per year, largely from natural sources present in the soil since the Earth's creation.[13] Today, those sources emit their radiation, for example, through the bricks and concrete from which our homes are built. The type of soil also matters: clay and loess contain more radioactivity than sand or peat.

Between countries, there are large differences in this natural background radiation. Outliers include Cornwall in the UK, and parts of Belgium, Finland and the Czech Republic, where large groups of people live with an annual natural radiation dose of 7 mSv.[14] The highest measurements occur among inhabitants of Ramsar, an Iranian city by the Caspian Sea, where thousands of people live with an average annual dose of 10 mSv, mainly because their houses are built with limestone from the surrounding area. In exceptional cases, annual doses reach 260 mSv. People living in such hotspots do not appear to present with health problems any more frequently than those in surrounding regions with lower radiation levels.[15]

Then there's radiation from space, from sources outside our solar system that collide with air molecules in the Earth's atmosphere. In the mountains, radiation from the cosmos is more intense than at sea level. Because we're closer to the cosmos on a plane, we're exposed to more radiation every time we fly. A return trip from London to Sydney yields about 0.1 mSv. The annual dose for pilots and flight attendants can be as high as 5 mSv.

Finally, there's the radiation involved in all kinds of medical procedures. A CT scan of the abdomen measures about 10 mSv. Four X-rays during a mammogram to detect abnormalities in the breasts measure about 0.6 mSv. When we have an X-ray taken at the dentist's office, it is about 0.003 mSv. Since American and Japanese patients use these kinds of scans more extensively than those in other countries, they get half of their total radiation dose through medical diagnoses and treatments. A rough, but controversial, maxim is that a dose of 1 mSv leads to an additional risk of one in 20,000, or 0.005 per cent, of developing and dying from cancer later in life.

According to UNSCEAR, 'sizeable population groups' are exposed to 10 to 20 mSv annually from a variety of radiation sources.[16] Studies suggest that our bodies can adapt well to those kinds of doses, although they could contribute to cancer.

Back to Chernobyl. How much radioactivity was released when the reactor exploded? How much radiation did people in and near Pripyat absorb? What about those further afield in Western Europe?

The figures get a bit murky here. After the accident, employees of the nuclear plant, along with firefighters, came into contact with several different types of radiation in various ways. Staff were covered in radioactive dust from the explosion and splashed with radioactive steam from ruptured pipes. Some stood ankle-deep in radioactive water; others breathed in radioactive dust. Some dust quickly dissipated, while other particles scorched their airways. There were workers who kept their overalls on all night, leaving their skin irradiated all that time.

In the case of local residents, it is equally difficult to determine the amount of radiation they received owing to the accident. Several factors will have had an impact: whether they opened their bedroom window that night, whether they went outside or stayed in the day after the accident, whether they washed their hair and clothes, or whether they were outside when it started raining.

Nevertheless, UNSCEAR has estimated that, on average, residents of the most affected areas in Ukraine, Belarus and Russia would have received between 10 and 30 mSv – not annually, but added together in the 20 years to 2005. By comparison, countless residents from Belgium to Brazil, from Finland to India are exposed to much more than 100, even as much as 200 mSv in 20 years of natural background radiation alone.[17]

UNSCEAR mentions that among the hundreds of thousands of clean-up workers in particular, there are uncertainties about the dose they received. For them, the total radiation dose in the years up to 1990 was estimated at an average of 120 mSv. Some 85 per cent of these workers received a dose somewhere between 20 and 500 mSv, with outliers above 1,000. These estimates are on the high side, according to the researchers themselves.

If UNSCEAR widens the circle to the six million residents of the most contaminated areas in the former Soviet Union, then their average radiation dose in the 20 years since the accident is estimated at 9 mSv. If it expands the circle to the nearly 100 million people living in Ukraine, Russia and Belarus at the time, the estimated average dose is 1.3 mSv. UNSCEAR speaks of 'an insignificant increase over the dose due to background radiation over the same period'.[18]

And the remaining hundreds of millions across Europe? Their total dose of radiation from Chernobyl, over a lifetime, is no more than what nature imperceptibly administers to them every year. In the UK, the Chernobyl accident annually contributes to a total of 0.0054 mSv in additional radiation.[19] That number includes fallout from atomic tests.

The huge gap between the perception and reality of radiation indicates that, as a species, humans have great difficulty in rationally assessing risk. We imagine flying is more dangerous than driving a car. We fear a terrorist attack, even though a family argument is more likely to be fatal to us. Basic knowledge of statistical probability changes little, because we humans are selective in how we process information. Even the mind of the smartest person can be held hostage to incorrect assumptions and irrational fears.

Exposure to ionising radiation leads to illness or death only in highly exceptional situations, but the discomfort is always there. When we are having a scan in hospital, we consciously choose exposure to radiation. That's not the case when an accident occurs in a nuclear plant. Do we know for sure that radioactivity is not escaping from there? Experts may be able to measure radiation accurately, but can they be trusted? If the effects of radiation ever do occur, it will be much later, insidiously. The damage is difficult, often impossible, to determine. It is impossible to perceive.

That doesn't sit well with us.

However, there are a few rules of thumb. The body can cope well with radiation. It will tend to repair radiation damage by itself, as it does with all tissue damage. Significant, irreversible damage only results from a dose level that we are unlikely to contract. Radiation doses reach deadly levels only in exceptional circumstances. And when these occur, death usually comes later in life, after a common illness – cancer – for which there can be all kinds of causes.

All the while, the danger remains hidden from our senses. That sounds scary, but we can also look at it differently. Perhaps, as a radiology expert once remarked, there is a good explanation for why we do not have a sensory organ with which to register radiation: we have no reason to.[20]

And yet, that is not the whole story. Health is more than the matter of a functioning body and a beating heart. The radiation released in a nuclear plant accident may not do much to our physical body, but it can significantly impact our mental health. This became apparent when a trembling of the ground signalled the beginning of a new nuclear nightmare in Japan...

CHAPTER SIX

Exodus

What should we (not) do after a nuclear accident?

'But as I travell'd hither through the land,
I find the people strangely fantasied;
Possess'd with rumours, full of idle dreams.
Not knowing what they fear, but full of fear'

— William Shakespeare, *King John*

Satoru Yamauchi misses his noodle restaurant. Returning after many long years, he acknowledges it meant everything to him. 'It was my life,' he tells a reporter of the French press agency AFP.[1] His voice cracks.

It will be hard to start all over again. Growing rice or picking wild plants in this area is now forbidden. How will he prepare his famous tempura with seasonal vegetables? And for whom? Yamauchi is yet to see many potential customers in Naraha, the first village in the province of Fukushima to be declared habitable again in 2015, four and a half years after radiation escaped from the nearby nuclear plant and some 160,000 people were forced to evacuate.

Yamauchi will never forget that disastrous Friday on 11 March 2011. In the early afternoon, he was working in his noodle restaurant when the ground began to tremble beneath his feet. It would be one of the world's worst earthquakes ever measured. Throughout eastern Japan, buildings shook and collapsed. Some people were trapped, others crushed. Gas and power lines broke. Fires started.

Then Yamauchi heard a warning: a tsunami was coming. He made his way out and dashed up the hill. The first waves rushing over the land were 10m (33ft) high, much higher than the alarm had announced. The highest was nearly 40m (130ft). Cars, houses and entire villages were washed away.

The force of the natural violence on that cold day is beyond comprehension. Japan's main island shifted a few metres to the east. The vibration of the seabed reached as far as Antarctica, where giant chunks of ice broke off. All the way on the other side of the Pacific Ocean, off the coast of Chile, the waves were still 2m (over 6ft) high. The death toll was eventually determined to be almost 20,000, plus several thousand people were still missing.

For Satoru Yamauchi, the damage was minor, or so it seemed. He, his wife and their four children were unharmed. Their house on the hillside was not swept away. Their dog was doing fine. Yamauchi's family was relieved to find themselves among the lucky ones.

But soon enough, their luck changed. Something was wrong at the Daiichi nuclear plant in Okuma, 20km (12 miles) away. Sensors had detected the coming of an earthquake and the reactors had shut down automatically. However, there was a problem with cooling the fuel rods in the reactor cores. The power needed to pump water around failed and the emergency diesel generators in the basement weren't working either, owing to water damage. The temperature inside the reactors was rising.

When Prime Minister Naoto Kan appeared on TV at around 5 p.m, he did not share this information. In a brief statement, he expressed his condolences to compatriots who had been affected by the tsunami. Then suddenly, from out of nowhere, he said: 'As for our nuclear power facilities, a portion of them stopped their operations automatically. At present we have no reports of any radioactive materials or otherwise affecting the surrounding areas.'[2] He asked everyone to remain calm.

Behind the scenes, Kan himself was anything but calm. The head of government, already plagued by political affairs, shouted and snarled at his staff. As award-winning journalist

Yoichi Funabashi recounts in his excellent 2021 book *Meltdown: Inside the Fukushima Nuclear Crisis*, Kan feared the consequences of an overheated nuclear reactor and kept saying, to no one in particular, sometimes loudly, sometimes muttering: 'It's the same thing as Chernobyl! It'll be just like Chernobyl.'[3]

While some prayed for the nuclear reactor to cool down quickly, a close associate wrote in a memo: 'It's Kan who needs cooling down.'[4]

Later that evening, he declared a nuclear emergency. Evacuation orders followed in quick succession. First a 2km (1.2-mile) radius around the nuclear plant, then 10km (6 miles), then 20km (12 miles). Now the Yamauchis in Naraha had to move as well, straight away. Buses showed up, some with squealing tyres, and everyone squeezed in. This is how Satoru Yamauchi ended up in a shelter. He made himself useful in the soup kitchen.

The days were full of dread. Workers at the Tokyo Electric Power Company (TEPCO), the owner of the nuclear plant, worked frantically to bring the situation in Daiichi under control. In the absence of power to keep the cooling system functioning, the water might eventually turn into hydrogen, which could explode. Indeed, in the following days, there were three hydrogen explosions in the reactor buildings. Everyone saw the footage. Everyone was shocked.

This time it was the deputy chief cabinet secretary who said: 'Isn't that an explosion like the one at Chernobyl? Isn't the same thing happening that happened at Chernobyl?'[5]

Rumours abounded. Emperor Akihito was said to have fled the capital, or even the country. There was mention of a mushroom cloud. A mushroom cloud? No one in Japan had forgotten the horror of the atomic bomb. *Let's get out of here!* The roads were jammed with cars. People stuck in traffic had escaped the rushing water, but now were wondering how to outrun the mysterious poison in the air. It was said that most of the radiation was blown towards the sea. But what if the wind turned? What if it started to rain or snow?

Satoru Yamauchi, too, was in doubt. His children begged him to move to Tokyo. They said they didn't want to die from radiation.[6] And there, in Tokyo, at a safe 200km (125 miles) from the nuclear plant that would dominate world news for weeks, their problems began. Feelings of depression and a lack of purpose bubbled up and wouldn't budge. At school, the children were bullied and excluded; the other children said they were radioactive. The family also faced financial problems, despite receiving the government's monthly allowance provided to all those who had to flee their homes.

Years later, back in Naraha, Yamauchi acknowledges that his family was psychologically wrecked. He now takes pills for high blood pressure. Like tens of thousands of others, he tried to build a life elsewhere, and is now full of doubts upon his return. 'I want my old life back,' he tells AFP, 'but I don't think it's possible here.'[7] For him and his family, it feels like they are living with a death sentence, marked by the radioactive cloud hovering over them.

The return of the Yamauchis and their fellow villagers to Naraha in 2015 was made possible due to the government finally lifting the evacuation order. The abandoned areas have now been sufficiently cleaned. The topsoil has been scraped off and disposed of in bags, the earth shovelled over. Houses, offices and streets have been rinsed clean. Leaves have been removed from the trees. The cost for recovery in the province: about 7 trillion yen (more than £40 billion, €45 billion or $50 billion). And the costs are still spiralling.

However, while the invisible danger has been brushed away, something just as imperceptible has taken its place: suspicion. Is it really safe? Are there really no longer any health risks? What's up with those Geiger counters the government distributed to all residents so they can measure the radiation level themselves? Have they been tampered with? After all, looking at those meters, nothing much seems to be going on. Were their leaders hiding something?

Satoru Yamauchi cannot feel relieved. 'There is nothing good about going back.'

If – yes, *if* – nuclear power has a future, much will depend on how we react after a nuclear accident. What happened in Fukushima does not bode well. Naoto Kan got it right: Fukushima became 'just like Chernobyl'. But not quite in the way he thought...

Just before a series of events to commemorate the 25th anniversary of the Chernobyl disaster, UNSCEAR published a report with findings on the accident's long-term health effects.[8] It came out 11 days before the tsunami hit Japan's east coast. Had Kan or one of his advisers read the report, they would have known not only that the increase in cancer at Chernobyl was very limited – hardly measurable in a large population – but that there was serious psychological damage among survivors caused by the fear of radiation.

People from the Pripyat area, the authors argued, were much more likely to suffer from stress, excessive anxiety and depression. This affected their behaviour: they drank more, smoked more, ate less healthily, lived more recklessly. The main public health impact of the world's biggest nuclear disaster was not physical, but mental: anxiety.

This conclusion was not new. In 2006, the Chernobyl Forum, the science-led collaboration between various UN agencies and national governments, found that many people from the region suffered from an accumulation of complaints. The authors talk of 'an exaggerated sense of the dangers to health of exposure to radiation', and of 'a widespread belief that exposed people are in some way condemned to a shorter life expectancy'.[9] With just about every cough, they would think: *is this because of the radiation? Is the deterioration starting now?*

Despite the facts, people were convinced that all kinds of diseases would now occur more frequently, and that these must have something to do with radiation.

Louisa Vinton, programme manager at the Chernobyl Forum, once observed: 'Fear of radiation is a far more important health threat than radiation itself.'[10]

After the power failed at the Daiichi nuclear plant in Fukushima, it was soon clear that radiation levels were much lower in the surrounding area than they had been at Chernobyl; nobody was hospitalised with acute radiation sickness. After the crash course in the previous chapter, we now know all about the millisievert. So what kinds of figures are we talking about for Fukushima and its surroundings? Let's turn once again to UNSCEAR.[11]

In the first year, evacuated residents were on average exposed to less than 6 mSv. That's lower than exposure from a pelvic CT scan. Immediately outside the evacuation zone, people were exposed to less than 4 mSv. That's pretty much the dose a full-time flight attendant absorbs every year. In the days following the accident, crowds of expats who wanted to be on the safe side flew home from Tokyo, surrounding themselves with more radiation, high up in the atmosphere, than if they had stayed put.

And so it's no wonder that UNSCEAR reports, in its two studies on Fukushima published in 2014 and 2022, that there is no discernible increase in health effects linked to radiation. There has been no increase in birth defects or heart attacks, nor thyroid cancer, leukaemia, breast cancer, colon cancer or any other cancer that one might attribute to radiation exposure from the Daiichi nuclear plant accident. Nor does the UN Commission expect such an increase. Indeed, the radiation dose incurred by the population is considered to be 'low or very low'.[12]

Of course, the dose is higher for those who worked at the damaged nuclear plant. Of the over 20,000 people working there in the year and a half after the accident, six exceeded the 250 mSv limit set for emergency workers. Excluding these six, the average annual dose in the first year hovered around 13 mSv – the equivalent of one chest CT scan and one of the spine. By the second year, the dose had more than halved.

Among aid workers, UNSCEAR doesn't expect to see an increase in health effects related to radiation either. Perhaps some will suffer from cataracts, since not everyone followed regulations and shielded their faces properly while working. Less than 200 workers were exposed to more than 100 mSv. They now have a slightly higher chance of developing cancer later in life – from a previous risk of, say, 30 per cent to now 31 per cent. Given the small group of people involved and the relatively high probability of getting cancer anyway, they will not be noticed in the statistics.

The scientifically established number of radiation deaths around Fukushima is zero.

However, a Japanese court ruling in 2018 put the number of radiation deaths at one. This was a former employee who had worked in several nuclear plants since 1980, the last few years in Fukushima. According to a provision in the law, that man's fatal lung cancer could theoretically be attributed to the radiation he contracted there.[13] So: zero in the world of science, one in the world of politics.

By contrast, the evacuation itself led to many fatalities, and these are not disputed. The UNSCEAR report confirms that in the chaos during and immediately after the evacuation, dozens of hospital patients died. Hundreds of elderly people passed away as a result of the abrupt move. Some were accidentally left in their rooms, where they either forgot or were unable to take their medicine. They perished as a result. According to the Japanese government's definition, produced nine years after the Fukushima accident the number of deaths owing to the evacuation and associated stress was put at 2,313.[14]

In 2016, the World Health Organization concluded that the Chernobyl disaster's greatest impact on public health was its 'psycho-social impact'.[15] Similarly, the Fukushima accident resulted in higher incidences of depression, post-traumatic stress and alcoholism among survivors.

The impact was especially evident among those people – both adults and children – who had been evacuated. They were more likely to suffer from obesity, diabetes and high blood pressure than those who had not been moved out. UNSCEAR offers an explanation: they lost their homes, their jobs and their sense of community.

Once more, similar conclusions emerged from earlier research on Chernobyl, where 350,000 people had to relocate. That experience was 'deeply traumatic', in the words of the Chernobyl Forum.[16] 'Many are unemployed and believe they are without a place in society and have little control over their own lives.'[17] There was said to be a 'paralysing fatalism' among evacuees and residents of areas around the nuclear plant after the events at Chernobyl.[18]

The few people who returned to their homes after some time, against the rules (although the authorities turned a blind eye), were considerably happier than those who stayed away. Many journalists discovered this when visiting the Pripyat area years after the disaster. These *samosely* – the Russian name for the people who settled illegally in the inaccessible 'exclusion zone' – often grew food in their gardens, collected herbs from the forest, hunted wild animals and fetched water from local wells. Comprising many hundreds, even thousands, they formed a stiff-necked, now largely extinct tribe. They often appeared healthier and happier than their former compatriots, who could not adjust to life in a different environment, free of radiation but full of worries.

Those who were evacuated after the explosion at Chernobyl suffered the same mental problems, such as stress and depression, observed in those who experienced the atomic bombing of Hiroshima and Nagasaki. But where the Japanese saw themselves as *survivors*, the Soviet people saw themselves as *victims*. That feeling was reinforced by the politicians who offered financial compensation to some seven million people, plus an endless list of preferential treatments, from free medicines and dental care to discounts on public transport and holiday vouchers. Many came to see themselves, reports the Chernobyl Forum, as 'helpless, weak and lacking control over their future'.[19]

Those forced to leave their homes after a nuclear accident, whether at Chernobyl or Fukushima, suffered from a stigma: they were excluded and shunned in their new surroundings, sometimes even by their own family. Placement of the evacuees into other communities introduced tensions, where there was a perception that they were taking homes and jobs. Many chose not to reveal where they were from, instead making up a family history.

Social exclusion of those who have been in contact with radiation is a well-known phenomenon. The crew of the Japanese fishing boat who all fell ill from fallout after a US nuclear test in 1954 (see Chapter 2) were shunned for a long time after their recovery. The *hibakusha*, survivors of the atomic bombings of Hiroshima and Nagasaki, had great difficulty finding marriage partners. Perhaps, it was thought, they gave off radiation. Perhaps their children would be infected and deformed.

Fear of contamination is not confined to Japan. After the Chernobyl accident, some healthcare personnel were apprehensive about treating firefighters and employees from the nuclear plant because radiation sickness was said to be contagious. Local residents who had to evacuate were shunned by their new neighbours. Parents forbade their children to play with their peers from Pripyat or sit next to them in class.

It's an irresistible zombie logic – those who are poisoned become poisonous themselves. But radiation actually doesn't spread easily, and certainly not once the person exposed has taken a shower and washed their clothes.

Such misconceptions are clearly harmful. For instance, Kai Watanabe, a twenty-something who signed on for clean-up work at the Daiichi nuclear plant, believes he will probably never be able to get married. After all, if he ever meets the woman of his dreams, he will one day have to confess to her that he worked at the plant. 'Let's say I tell a woman about my past, that I've absorbed all this radiation and may get sick or father children that are deformed so we shouldn't have children,' he says in *Strong in the Rain*, a book with moving personal stories of survivors. 'Is there a woman out there who would accept me?'[20]

Studies also show that more than 40 per cent of young mothers in Fukushima have strong feelings of anxiety because of the stigma, and that young women have negative feelings about the prospect of pregnancy.[21]

If the tsunami left an upsetting trail of devastation, the nuclear accident left a devastating trail of upset. In our minds, these effects soon became mixed up, just as atomic bombs and nuclear plants were once confused.

It must have been hard to accept that a modern, advanced society could be so overwhelmed by nature, as it was in the earthquake and the tsunami. Politicians realised that although they had not been able to protect their citizens from a tsunami, they could still safeguard them from being exposed to radiation from a damaged nuclear plant. It suited them politically that citizens began to worry about all the terrible diseases they might contract due to the nuclear accident.

Some wondered why so much attention was being paid to the nuclear plant which, like so many other buildings, was not adequately protected from the enormous waves – there hadn't even been a single fatality there, unlike all those apartments and offices where tens of thousands were killed instantly by the tsunami. Forget the nuclear plant, why were the Japanese people not better protected from the waves?

It's a cruel irony that the Fukushima accident should have come at such an unfortunate time. In politics, the threat posed by climate change had finally begun to sink in. Slowly, the realisation dawned that a source of reliable, round-the-clock, carbon-free energy would be useful. Didn't nuclear energy deserve another chance? Chernobyl was already so long ago. Surely reactors had got much safer since then?

Industry lobbyists were already speaking of a 'nuclear renaissance'. However, this was somewhat premature; global electricity production from nuclear plants was no longer increasing, but fluctuating. More nuclear plants were closed than opened.

But a change was in the air. And at the vanguard were, somewhat surprisingly, thought leaders in the environmental movement. Stewart Brand, a pioneer of the 1960s counterculture, and Stephen Tindale, a former chairman of Greenpeace, had opposed nuclear power and while they remained enthusiastic about the potential of solar panels and wind turbines, they now realised, in Tindale's words, 'that renewable energy cannot expand quickly enough to phase out fossil fuels and protect the climate'.[22] They acknowledged their misjudgement and showed a willingness to accept nuclear power as a viable supplement to renewables.

Even the events at Fukushima didn't necessarily put an end to that willingness. Indeed, some used the accident as an example for arguing in favour of nuclear power. Ten days after the accident, George Monbiot, one of the intellectual forerunners of the green movement, wrote in the *Guardian*:

> *A crappy old plant with inadequate safety features was hit by a monster earthquake and a vast tsunami. The electricity supply failed, knocking out the cooling system. The reactors began to explode and melt down … Yet, as far as we know, no one has yet received a lethal dose of radiation … The crisis at Fukushima has converted me to the cause of nuclear power.*[23]

Most drew a different lesson. According to them, it had been proved that an accident could not be ruled out and that nuclear power is therefore unacceptable. Satoru Yamauchi, the noodle restaurant owner in Naraha, is adamant. In an interview with the *Financial Times*, he said, 'There's absolutely no need for nuclear power. With just one mistake, terrible things happen.'[24]

The leader of Europe's largest economy sided with the chef. Chancellor Angela Merkel immediately closed seven nuclear plants in Germany. Encouraged by a mass protest, in which 200,000 Germans took to the streets, and referring to an old promise – made when the Greens formed the government together with the Social Democrats – she had decided to phase out nuclear power. At the time, nuclear power accounted for some 25 per cent of all electricity in Germany. The last German nuclear plant would close by 2022.*

Germany was not alone. The Swiss government decided not to build any more nuclear plants. In a referendum, the Italian people spoke out, once more, against building nuclear plants within their nation's borders, choosing instead to remain a leading importer of nuclear power from French and Swiss facilities. South Korean and Taiwanese authorities also referred to Fukushima when announcing nuclear power would be phased out.

Japan itself decided all nuclear plants must close down. The island nation that once opted for nuclear power, because it had exhausted its own coal reserves and wanted to avoid dependence on other countries, closed dozens of nuclear reactors. Thus, Japan became one of the world's largest importers of natural gas and coal.

So much for the nuclear renaissance. As it happened, it wasn't the health of the Japanese people that took a hit from the events at Fukushima, but the nuclear industry.

Japan's nuclear industry had it coming. For instance, nuclear companies maintained a close relationship with the regulator. They assigned jobs to one another. The power company TEPCO had been advised to raise the sea wall of

---

* The last nuclear plant in Germany closed on 15 April 2023 – an extension having been necessary due to the energy crisis caused by Russia's invasion of Ukraine.

its Daiichi nuclear plant, but then did nothing and got away with it – until the tsunami.

Japan's nuclear clique had proved complacent, even though there had already been fatal accidents: in 1999, two radiation deaths at a uranium processing plant in Tokaimura, and in 2004, five fatalities resulting from a hot steam leak in the turbine building of a nuclear plant in Mihama. In both cases, safety procedures had not been followed. Staff, including managers, had been inadequately trained.

Politicians were equally ill-prepared for an accident. There was the occasional mandatory drill, during which role-plays prepared officials for an emergency at a nuclear plant. There had been one five months before the ground started shaking on 11 March 2011, as Funabashi recounts in *Meltdown*. Then too, Prime Minister Naoto Kan 'declared' a nuclear emergency. Sitting at a conference table with a few ministers, he seemed uninterested, reading text from a stack of papers. *What a waste of time*, thought one of the ministers. After an hour, he snapped at Kan: 'This serves no purpose.'[25]

Clearly, anyone who believes everything at their nuclear plants is just fine, as the Soviets and the Japanese had thought, is at a disadvantage. Safety culture fails when nobody is concerned about safety. In retrospect, this is an easy observation to make. Such problems should be identified and addressed with foresight, not hindsight.

In both politics and the nuclear industry, there can be a worrying surplus of trust; in society, there has been a serious deficit. This is not surprising. Inside and outside Japan, there is deep suspicion of the companies and organisations within the nuclear power sector. TEPCO did nothing to correct that image, staying quiet in the hours following the accident in Fukushima. Its top executives couldn't be reached; it has been suggested they were hoping to shift responsibility for what happened at the plant to the government.

Whatever TEPCO might have said, nobody would have bought it. It would have always sounded like an attempt to cover up the truth. To a journalist's simple question of whether or not there had been a meltdown, TEPCO's spokesperson replied that there was 'no evidence to specifically assert or determine either way'.[26]

But as anyone knows, leaving any room for ambiguity only leads to more questions and more concerns. Elsewhere, a Japanese government spokesperson emphasised that there was 'no immediate impact' on health.[27] It was meant to reassure, but anyone hearing that might think: *Ah, so the health impact will come later!*

So when Rafael Mariano Grossi, Director General of the International Atomic Energy Agency, was questioned about Fukushima on a stage at the 2021 UN Climate Change Conference in Glasgow and said that nobody had died from radiation, a sceptical audience began to chuckle. 'I don't know why you're laughing,' Grossi responded in surprise. 'It's a fact.'[28]

A fact it is, indeed. But when facts run counter to our thinking, things can get tricky. We have made nuclear power a spectre for so long that deep down we are convinced any accident in a nuclear plant must be of apocalyptic proportions. But the mundane reality is nothing like the wild fantasies we have put into our heads.

Fukushima was the first nuclear accident to occur in a modern society with 24/7 rolling news. Back in 1986, there were no images or footage of Chernobyl. Journalists encountered a veil of secrecy. But now, 25 years later, the moment had finally arrived! As if the natural disasters of the earthquake and subsequent tsunami were not enough, there had to be a nuclear catastrophe too.

Anyone re-reading news articles published in the wake of Fukushima will be struck by the overzealous reports of constantly lurking danger. The reactors were spewing deadly radiation. Radioactivity kept rising, well above safe limits. A disaster was inevitable. The population was doomed. The Japanese people would be felled by cancer in droves, emergency workers in and around the nuclear plant faced certain death, and the political and industrial authorities were hiding the terrible truth.

Experts with more nuanced analyses also featured. 'No Chernobyl is possible at a light water reactor,' explained a Japanese professor to press agency Reuters.[29] 'Loss of coolant means a temperature rise, but it also will stop the reaction.'

Yet such words proved to have little appeal; it was the stories featuring experts who believed current events in Fukushima were 'worse than Chernobyl' that were being read and shared.[30] One, Arnold Gundersen, a disgruntled former employee of the nuclear industry, told Al-Jazeera that we were witnessing 'the biggest industrial catastrophe in the history of mankind'.[31]

Editorial choices can lead to a curious cycle. First, journalists help to spread panic. Then, they turn the panic itself into news. Take for example the coverage in the *Sun*, the UK's biggest-selling national newspaper. A few days after the devastating tsunami, the *Sun*'s editors put the famous yellow-and-black symbol for radioactivity on their front page, alongside a headline screaming: 'Exodus from Tokyo – 1000s flee poison cloud'.[32]

No explanation was given in the paper as to how exactly such a 'poison cloud' could threaten Tokyo's 13 million inhabitants; even in the most pitch-black, worst-case scenario, no such thing was possible. However, the article stated that radiation around Fukushima was already approaching the level where people were vomiting uncontrollably, their hair was falling out and cancer rates were skyrocketing. *Panic!*

The next day, the *Sun* ran an op-ed by Brian Cox, the pop musician turned professor of particle physics and host of multiple popular science TV shows. Cox acknowledged that while damage to a nuclear reactor sounds scary, such a reactor cannot explode like an atomic bomb, and that the Japanese plant was not like the one at Chernobyl. There were only small amounts of nuclear material in the steam released into the air, Cox explained. 'The levels of radiation released in this way are very small – probably about the same as you would expect on a long-distance transatlantic flight.'[33] *Don't panic!*

But then, two days later, the *Sun* published a report by a British expat: 'My nightmare trapped in City of Ghosts'.[34] She was talking about Tokyo. She wrote that radiation levels had already increased tenfold. The city streets were grimly empty. It was like a zombie movie, she said. 'What if, every day, radiation continues to double?'

Fear sells, and after a nuclear accident, fear abounds. In the 1979 Three Mile Island accident, nobody was injured or fell ill; the radiation released to nearby residents amounted to 0.08 mSv, or several panoramic dental X-rays. Still, Walter Cronkite, America's best-known news anchor, addressed the nation with these words: 'The world has never known a day quite like today. It faced the considerable uncertainties and dangers of the worst nuclear power plant accident of the atomic age. And the horror tonight is that it could get much worse.'[35] When it soon turned out things weren't that bad, the tone barely changed.

A nuclear accident is a goldmine for the news industry. After Fukushima, journalists stuck to their disaster-laden script. Modern society's free press treated the facts as lightly as a Soviet state broadcaster.

In the Netherlands, journalists looking for context turned to Wim Turkenburg. In the early 1970s, Turkenburg teamed up with an anthroposophical teacher and other kindred spirits (see Chapter 3) to advise the government against building nuclear plants, before co-founding and chairing a think tank which, according to its own website, is 'closely linked' to the rise of the Dutch anti-nuclear movement.[36] After Fukushima, Turkenburg became a media personality – not as a co-founder of the anti-nuclear movement, but as a professor of science, technology and society at Utrecht University.

On the same day that Professor Cox provided a nuanced explanation in the *Sun*, Professor Turkenburg was a guest on Dutch public broadcaster NOS. 'Energy expert', it read at the bottom of the screen. Turkenburg called the situation in Fukushima 'extraordinarily serious'.[37] After all, he said, a radiation level of 4,000 millisieverts per hour had been measured somewhere, and if we consider that an ordinary citizen is only allowed 1 extra millisievert per year, then, yes, it is 'extraordinarily worrying' if somebody is exposed to those levels of radiation. According to Turkenburg, we could assume that employees in Japan were contracting 'all kinds of radiation diseases'.

But, the news anchor countered, aren't the Japanese government's announcements somewhat reassuring?

Not so, Turkenburg maintained, saying he had received very different information 'through other channels', and was finding it all 'very confusing'.

So what was his advice?

'Evacuate. Fifty to eighty kilometres [30 to 50 miles].'

Should the staff at the nuclear plant also leave?

The 'energy expert' searched for words. 'Well,' he began, 'one could also say: *I will sacrifice those people.*'

In addition to commentary from the experts, the stories coming out of Fukushima were all so saddening that there was barely a glimmer of hope. On 24 March 2011, two weeks after the accident, Hisashi Tarukawa, a farmer from Sukagawa, 60km (almost 40 miles) from the nuclear plant, heard he could no longer sell his rice, cabbage and other crops because of the increased radiation. He hanged himself from a tree in his field. His son found him.[38]

Tarukawa may have been the first in a long series of suicides linked to events at the Daiichi nuclear plant, including Hamako Watanabe, who had to leave her home. In June 2011, she returned, doused herself with petrol and set herself on fire. After she went missing, her husband discovered her charred remains at their chicken farm.[39]

A dairy farmer left a message for those left behind: 'If only there wasn't a nuclear power plant.'[40]

A 93-year-old woman, wrote in the note she left: 'I would only slow you down. I will evacuate to the grave.'[41]

By 2017, 99 suicides had been recorded relating to
Fukushima.[42]

After Three Mile Island, Chernobyl and Fukushima, we are
now more aware of what can go wrong after a nuclear
accident. Even at modern plants, radiation can escape. The
radiation will rapidly decrease; the substances with the highest
radioactivity decay in a matter of seconds, minutes, hours or
maybe days. For employees who remain at the plant to get
things under control, radiation may have an impact later in
life. However, the effect of radiation on public health is so
small that it cannot be measured.

With some straightforward advice for people in the region,
the response to the situation is quite manageable. Stay indoors.
Close the windows. Wash your clothes. Take a shower.

The government will have to take measures: start advising
whether and how to use iodine tablets (so that children in
particular can saturate the thyroid gland before radioactive
iodine from the nuclear plant accumulates there), monitor or
confiscate local dairy products for a month or two. That's
about it.

If this information is not enough to counter fears, maybe
it's time for exceptional measures. An international
commission of experts is on standby as an advisory body in
any nuclear accident, but should it perhaps wrest control of
crises from national authorities? And shouldn't those foreign
experts then take up residence in the villages surrounding the
nuclear plant, preferably with their families in tow, and while
we're at it, shouldn't we bring in the country's prime minister's
family as well? Could such symbolic gestures ever win back
the trust that the nuclear industry has long since lost?

We have seen how nuclear accidents generate a lot of
uncertainty, which can easily turn into fear. It takes little
more than overly firm politicians, failing communication
managers, excited journalists and confused experts to generate

a full-blown crisis. They fuel the persistent feeling that disaster strikes whenever something goes awry at a nuclear plant.

Of course, nuclear plants are not infallible, and we will have to learn to live with the fact that accidents cannot be ruled out. While it will take a whole lot more than a leaking pump, a crack in the concrete or an inattentive employee, the probability of an accidental release of radiation will never be zero.

When something goes wrong at a nuclear plant, things presumably go wrong at the newsrooms too. Such accidents will never stop catching the attention of journalists. Forget yet another gas explosion or flooded coal mine; it is precisely the rarity of nuclear accidents that makes them so newsworthy.

But what would help enormously is if editors were to check credentials when selecting experts. They do exist: competent experts with a good track record and respect from their academic peers. Those who were really paying attention after Fukushima would have found them. After readers of the *Sun* learned of that nervous expat's 'nightmare in Tokyo', they could have turned on Channel 4 and seen an interview with Geraldine Thomas, an expert on Chernobyl and cancer, and author of several scientific studies on radiation and health. She explained that the radiation in Fukushima did not seem too bad, that any health risk would be restricted to the unfortunate emergency workers at the nuclear plant, and that it would be enough to evacuate only those in the immediate area.

'One thing we should have learnt post-Chernobyl', Thomas said, 'is not to spread panic and make claims that turn out to be wrong. The psychological damage being done now to the Japanese is huge.'[43]

However, meanwhile, in the Netherlands, Wim Turkenburg was all over the news. His university even boasted on its website that Turkenburg had become 'the face and voice' of the Dutch public broadcaster NOS in the two weeks after the accident.[44] In fact, it had become 'impossible to imagine the NOS studio without him'.

Initially, the news desk had invited Tim van der Hagen to appear, a professor of nuclear reactor physics serving Delft University of Technology as dean of the faculty of Applied Sciences, as well as director of the Reactor Institute Delft, a research centre on radioactivity. But while his clear expertise resulted in informative television, the editors weren't happy. They'd seen the foreign news programmes that made the accident sound much more exciting.

And so the morning after the programme aired, Turkenburg's phone rang. It was the NOS evening news. They summarised the problem: they'd seen that Van der Hagen was 'very reassuring'. Turkenburg couldn't have agreed more. 'I had also heard him, and I was not happy with his statements,' Turkenburg confessed on his university's website. And so he was asked if he might want to come to the TV studio to educate the Dutch viewers on Fukushima. He considered it his duty.

In March 2021, as the media looked back in detail at the events of 10 years earlier, a researcher noticed something odd after studying 35 articles from a range of Dutch newspapers: almost all focused on the aftermath of the nuclear plant accident; the *real* disaster of the earthquake and the tsunami was covered 'at most in passing'.[45]

Journalists had spoken to residents who had been evacuated from the areas surrounding the nuclear plant. But not one had talked to those who had lost loved ones in the natural disaster. The stories centred on Fukushima, and not areas farther away where the tsunami had hit much harder. It's also worth noting that there are a number of nuclear plants closer to the epicentre of the earthquake, but nobody ever heard about them because those facilities had not failed.

'One-sided and misleading,' or so judged the media researcher, who must have been in an amiable mood.

In the festival of flaws, the grand prize went to... the NOS evening news. The anchor managed to mix up the sequence of events: 'Ten years ago the nuclear disaster in Fukushima, Japan, was followed by a tsunami.'[46]

It was, of all media outlets, *De Telegraaf*, known as the most sensationalist tabloid in the Netherlands, that offered context and perspective by interviewing Geraldine Thomas. She aptly summarised the scientific consensus: 'Nobody died from radiation released in Fukushima, and nobody will die from it.'[47]

Lastly, the evacuation. In the event of 'emergency exposure situations', the International Commission on Radiological Protection recommends protective measures for 'reference levels' in the band of 20 to 100 mSv, adding that a dose rising towards 100 mSv will 'almost always' justify measures.[48] The Japanese government chose the lower end of the spectrum and decided to evacuate areas where the annual dose of radiation in the air, just above the ground, amounts to 20 mSv: a level you would be exposed to if you were outside 24 hours a day. Now those in power could *do something*. They started moving out hordes of people.

The appetite for evacuation was insatiable. Even from areas that remained below the 20 mSv level, residents were forced to relocate. As late as June 2011, when it was long since clear that the darkest scenario had not materialised and the radiation level was already falling on its own, people were still forced to leave. Not all were able to stay in the Fukushima province. Some had to move again and again, endlessly dragging themselves from shelters to temporary housing.

Vast areas were declared uninhabitable owing to a radiation level that countless people in Finland, the Czech Republic, England, Brazil, China, India, Australia, Iran and numerous other countries live with on a daily basis without any adverse health effects. If it were really only safe to live below the new radiation level at Fukushima, many millions of people around the world would have to move.

According to conventional models, a little extra radiation exposure can slightly increase the chances of one day dying of cancer. From 100 mSv upwards, 1 extra mSv is estimated to mean an extra risk of 0.005 per cent.

Probability is a tricky concept for many people, so instead experts prefer to translate the calculations into more illustrative terms, namely how many days on average a person shaves off their life through, say, smoking, eating fatty foods or living in a city with a lot of air pollution. The same can be done for exposure to a certain dose of radiation. It's equally possible to calculate how many days a person's life is *extended* by avoiding radiation through relocation.

For example, the inhabitants of Tomioka, the town near the nuclear plant that experienced the highest radiation levels, but whose residents were less exposed to radiation thanks to relocation, extended their lives by two months, three weeks and one day. For those leaving other areas, it was much less; thanks to their evacuation from Naraha, noodle restaurant owner Satoru Yamauchi and his fellow villagers extended their lives by no more than a few days.

These figures come from Philip Thomas, a professor of risk management at the University of Bristol. He led a study conducted by a number of universities. The conclusion: the evacuation at Fukushima was excessive.[49] He wants to prevent the Japanese government's response from becoming the prevailing policy choice after a nuclear accident.

According to Thomas (no relation to Geraldine), it is justifiable to evacuate the immediate area after a nuclear accident – say, a radius of a few kilometres around the plant. But after a few days or at most a few weeks, everyone should be able to either return or relocate. A short evacuation has the least impact on well-being – the longer it lasts, the greater the disruption.

The International Commission on Radiation Protection (ICRP) has reported it would not recommend evacuation 'for a period of longer than a few weeks'.[50] The IAEA states it should be a week at most.[51]

So, what can be done about the impulse to run? Philip Thomas has a radical plan: provide information. With information – about what radiation is and what it does, about the doses and the millisieverts, about the effects on health and longevity – people can make their own choice as to whether to stay in the area. If they want to leave, there should be a financial settlement – not a monthly payment that could lead to dependency, but a decent, one-off compensation.

'With hindsight, we can say the evacuation was a mistake,' Thomas said in the *Financial Times*.[52] 'We would have recommended that nobody be evacuated.'

Nobody.

What should we do when it becomes apparent that the accident turned out differently from the way we all expected? What if we accept that 160,000 people were forced to move unnecessarily and that they have lived in uncertainty for years, that their lives have been severely disrupted, their health undermined? What if you were among the experts or journalists who appeared to 'know the truth' of what was going on? What if you were a member of government, or even the prime minister, when all of this happened?

Naoto Kan said in 2016, five years after stepping down as leader, that the situation in Fukushima was so dire that he had considered imposing martial law. Interviewed by *The Telegraph*, he said, 'The future existence of Japan as a whole was at stake.'[53] Kan considered evacuating a 250km (150-mile) radius, including the metropolis of Tokyo. The moment almost came when he had no choice. It was only a hair's breadth away, Kan said, and it was all thanks to the courage of his people who risked their lives to contain the nuclear plant – indeed, to save their country. Indeed, the way in which he had averted an almost inevitable catastrophe was truly ingenious.

Nuclear power? Naoto Kan wanted no more of it. 'Next time, we might not be so lucky.'

Without a doubt, people like noodle chef Satoru Yamauchi have already felt how 'lucky' they have been.

PART THREE
HUH?!

CHAPTER SEVEN

Peace With the Atomic Bomb

Have nuclear weapons really made the world less safe?

'Si vis pacem para bellum.'
(If you want peace, prepare for war.)

— Ancient Roman proverb

At the turn of the twentieth century, H.G. Wells was the undisputed master of science fiction, a worthy successor to Jules Verne. His books entertained millions of fans. Wells' stories are full of fantasy about time travellers and invisible creatures, but above all full of hope for modern life, which would be made possible by technological advances.

But when European politics became dominated by the lust for power and the atmosphere turned grim, Wells felt overwhelmed by gloom. Science and technology, which he had so often praised, would surely mean that the next war would be more horrific than any war ever seen before.

Consider motorised aircraft. The Wright brothers had barely got an aeroplane off the ground before someone considered dropping bombs on the enemy from the air. In 1911, Giulio Gavotti, an Italian pilot fighting Ottoman Empire troops in Libya, placed three bombs at his feet and one in the pocket of his jacket. From his rickety little plane, he tossed them overboard.

My goodness, thought Herbert George Wells. *Where is this going…*

Wells retreated to a chateau in Switzerland and sat down to write his next novel, *The World Set Free*. It's not exactly an easy read. When the book was published in 1914, it received scathing reviews, if it was noticed at all. Yet this obscure volume is said to have predicted and influenced the course of the future more than any other of his works.

In *The World Set Free*, Wells outlines a world war, some time in the mid-twentieth century, where nations like the UK and France would face off against an alliance between Germany and Austria-Hungary. Both sides would possess a new weapon with destructive powers. Wells even came up with a name for it: the 'atomic bomb'.

Wells got the idea from a book by the English radiochemist Frederick Soddy about radium and the structure of the atom. Soddy recounted a conversation with his colleague, Ernest Rutherford. They observed the staggering amount of energy inside the atom and let their imaginations run wild. Consider the atom as a weapon. Whoever found the detonator, Soddy wrote, 'would indeed make this old world vanish in smoke'.[1]

Wells quoted liberally from Soddy's book, dedicated *The World Set Free* to it and then launched into a hotchpotch of pseudo-science. The novelist spoke of 'Carolinum', a radioactive element that can be cracked open to release energy. This is possible thanks to an imagined scientific discovery Wells places in 1933. A handbag with a lump of Carolinum contains enough energy to leave half a city in ruins. Packed into a bomb, pilots dump the stuff overboard. The inferno that ensues scorches anything nearby. Wells specifies that this bomb doesn't just go off once, *kaboom!* – no, this weapon 'continued a furious radiation of energy and nothing could arrest it'.[2]

In 1932, a copy of the book fell into the hands of – there he is again – Leó Szilárd. He was familiar with most of the works of Wells and had met the writer once at a dinner party in an unsuccessful attempt to secure copyrights for Central Europe. This book, he had missed. He read it carefully. A year later, Szilárd invented the nuclear chain reaction, the very scientific breakthrough Wells had predicted in his novel.

When stories about the Manhattan Project emerged after the war and Szilárd was asked who was truly the father of the atomic bomb, he would say: H.G. Wells.[3]

Many advocates of nuclear power shy away from the bomb. They have mentally drawn a firm line: nuclear plants good, nuclear weapons bad. Societal opinions on phasing out nuclear plants may diverge, but surely not on eliminating nuclear weapons? In actual fact, things are a bit more complicated – and a lot more interesting.

Whenever a president, dictator or terrorist leader flirts with nuclear weapons, international disapproval quickly follows. A line has been crossed. Everyone seems to agree, from peace activists and commentators to leading experts in the military or intelligence community: nuclear weapons are so devastating that they might destroy everything. An atomic bomb could spell the end of humanity.

None of that is new. It was also the story right after the atomic bomb was detonated over Hiroshima on 6 August 1945. Front pages spoke of a 'super-bomb'[4] and 'the most powerful bomb ever!'[5] President Truman boasted of 'a new and revolutionary increase in destruction'.[6] In a radio address, by way of surrender, Emperor Hirohito said: 'Should we continue to fight, it would not only result in an ultimate collapse and obliteration of the Japanese nation, but also it would lead to the total extinction of human civilisation.'[7] The Pope judged it 'the most terrible weapon which the human mind has conceived up to date'.[8]

In some cases, these words are literal echoes of what nuclear pioneers said in secret documents. In March 1940, two fugitive Jewish physicists informed the British government of the possibility of a 'super-bomb'.[9] Manhattan Project scientists talked about a 'new means of destruction'.[10] During the war, in a memo to the White House, the US secretary of war spoke of 'the most terrible weapon ever known in human history'. He warned that 'modern civilisation might be completely destroyed'.[11]

One thing they knew for sure: this weapon they were building would change everything. J. Robert Oppenheimer, scientific director of the Manhattan Project, said that after the first atomic test he and other observers 'knew the world wouldn't be the same'.[12] That wisdom was recognised in a newspaper headline: 'The bomb that has changed the world', only one day after it was introduced to the people of Hiroshima.[13]

H.G. Wells predicted something else that was readily accepted: every country would want a weapon like that. In *The World Set Free*, all nations were frantically making them so they could bomb the enemy to the ground before they themselves were wiped out.

Indeed, in the late 1940s, the Soviet Union quickly started producing and stockpiling nuclear weapons. Next came former colonial powers like Britain and France, followed by China. Piles and piles of nuclear weapons were designed, built, stored and tested. They were becoming ever more powerful. Some missiles with nuclear warheads could be launched from an underground storage facility or submarine and explode on the other side of the world, at the mere push of a button.

With the number of nuclear weapons on the increase, people started to believe that one was bound to be fired one day, thus unleashing a war. According to many, it was just a matter of time. The question was not *if*, but *when* and *by whom* these weapons would be used. Would anyone ever feel safe in a world where a single bomb can destroy so much?

The arms race was presented as rational, but was in fact absurd. If you fire nukes at an enemy that has them too, or at an ally of a country that has them, retaliation awaits. You can't really use the weapons, but you can't just give them up either.

And what happens when a nuclear war begins? Then, we were told, firestorms in our burning cities would pump smoke and soot into the stratosphere, where it would all clump together. The sun would be blocked for years and the Earth would get considerably colder. Crops would fail and the survivors would kill each other for what little food was left.

Everyone scared the hell out of everyone else. Government leaders were deterred from starting a war, but they were equally frightened of the enemy, so they wanted to better arm themselves. Traditionally, our rulers have sacrificed the poorest people in society in the theatre of war, but with the advent of the atomic bomb, their own survival was now at stake. Military strategists have always been trained to wage war and use weapons; now they had to master the art of *not* waging war and *not* using weapons.

Just as atomic scientists at the beginning of the twentieth century were engrossed in their dream world involving the possible interaction of invisible small particles, military strategists were now creating their fantasy of international relations with endless 'what-if' analyses. The only certainty everyone understood was this: if a war was waged with nuclear weapons, there'd be no hope of victory.

All the while, people were scared. Scared and prepared. Air raid sirens were tested. Some decided to build shelters, stocked with canned food and gallons of water.

This was not the public participation the bomb's developers had hoped for. Robert Oppenheimer was fearless in his warnings about the total power that nuclear weapons provide for any head of government who has them. Never before in history could a single person cause so much carnage with a simple order. That's why Oppenheimer advocated strong public engagement. In his view, people needed to gain a better understanding of the consequences of either having such weapons or giving up on them. Otherwise, the bomb would lead to paranoia and magical thinking.

But such understanding never came. By the 1960s, when the public was finally able to claim a role in the debate, the challenges had become so complex – even for those who called themselves experts – that they turned away from serious reflection. When people rallied in the streets, they simply shouted: 'Ban the bomb!' The slogan was not accompanied by arguments or considerations of what exactly would happen if the bomb were abandoned.

An international non-proliferation treaty, drafted in 1968, restricted the possession of nuclear weapons to a handful of countries. The treaty was celebrated as a victory. Only later came the realisation that this preserved the world order, in which the most powerful countries had the most powerful bombs while smaller countries were excluded from the privilege of ever having such weapons. How is that fair?

There was also little awareness of a rather simple truth: a world with nuclear weapons may not be safe, but neither is a world without them.

And then, something unexpected happened. Or rather, it didn't happen at all. *The nuclear weapons were not used.* Where war had been dreaded, peace emerged. The Cold War was a war in name only, because the Americans and the Soviets did not directly take up arms against each other. On the European continent, long since a battleground of war, an era of peace commenced.

The same thing happened elsewhere. Decades of violent unrest between India and Pakistan came to an end after both countries tested an atomic bomb. Two of the most notorious mass murderers in history, Mao Zedong and Joseph Stalin, had nuclear weapons but didn't use them. Was it self-control? Were the weapons enough to frighten others off? Were the leaders afraid to use them?

The world recorded fewer and fewer wars; wars counted fewer and fewer deaths.[14] In recent centuries, over so many generations, there has never been so few casualties of war. There is no misunderstanding about the beginning of that remarkably long and peaceful episode: it was after two atomic bombs were dropped on Japan and ended the most brutal war of all time.

H.G. Wells also predicted this. In *The World Set Free*, which is told from the point of view of someone living in the future,

Wells made it clear that at that time, in the heat of the moment, people were blind to something that was so obvious in retrospect, namely 'the rapidity with which war was becoming impossible'.[15]

'If we have them, why can't we use them?' That question, which referred to nuclear weapons, was reportedly posed by Donald Trump to a foreign affairs expert during his 2016 presidential campaign.[16] Throughout his tenure in the White House, nuclear weapons returned to the front pages. The news was not reassuring. A dispute with North Korea led to a back and forth involving much bellicose language. Trump signed an executive order for the development of new nuclear weapons.

Trump's policies marked a break with the recent past. His predecessor, Barack Obama, had stated 'clearly and with conviction America's commitment to seek the peace and security of a world without nuclear weapons'.[17] This provided widespread relief.

Just before Obama took office, six cruise missiles, each loaded with a W80-1 variable yield nuclear warhead, were accidentally loaded onto a bomber and transported from one state to another, after which they were taken to a storage bunker for 36 hours, left unguarded, without anyone having missed them. The incident suggested that nuclear weapons had lost their unique spell. Also, if we treat them so carelessly, hadn't we better get rid of them?

We dozed off because the Cold War was a distant memory, but Trump woke us up. Nuclear weapons? *Help, they're still here!*

Indeed, they are here. Today, nine countries around the world have nuclear weapons in huge numbers: just under 13,000.[18] Russia has the most, the United States comes a close second. The other countries follow at a distance. And while politicians often talk about disarmament, India, Israel, China and Russia are expanding their nuclear arsenals, for instance with longer-range missiles. The Americans, French and British are betting on innovation, developing small, 'tactical' variants. North Korea is believed to have a hydrogen bomb.[19]

And why stop there? In South Korea and Japan, politicians are arguing for their own production of nuclear weapons so they can keep the fickle dictator in Pyongyang in check, should their colleague in the White House not feel like it. Can Europe really expect the US to continue to protect its allies?

For years, the urgent threat of nuclear conflict has diminished, but concerns remain. And now that Vladimir Putin has shown his ambition to revive Russian influence in former imperial domains, anxieties have returned.

The more sinister the regime and the more unscrupulous the leader who threatens to use a nuclear weapon – from a Western perspective, of course – the greater the worry. Swift action is needed! Even the slightest suspicion of a secret nuclear weapons programme, as with Saddam Hussein's regime in Iraq in 2003, prompts diplomatic sanctions or military retaliation.

After al-Qaeda's attacks on the US on 11 September 2001, fears mounted that terrorists like Osama bin Laden would get hold of nuclear weapons. Perhaps they already had them. Hadn't some of the nukes gone missing when the Soviet Union disintegrated? Suppose they had fallen into the wrong hands…

Or was the problem that they were always in the wrong hands?

Now, suppose you want to get hold of nuclear weapons. You start by exploring the market. After all, nuclear weapons are for sale – everything is for sale. But where? Those few governments that own them won't just sell them to you. Why would they? By selling them, they give away power and make themselves vulnerable to sanctions once news of the transaction leaks. And when you, as the brand-new owner, fire the weapon, experts can determine where it came from based on the substances released into the atmosphere. There'd be a lot to explain. No government would want to take that risk. That's why nuclear weapons are so closely guarded.

But surely there is a black market? Yes, that is the rumour, especially in Eastern Europe. The question is whether weapons sold on the black market would actually work. A nuclear weapon requires specialist maintenance. If that's neglected, it automatically becomes a dud. Besides, nuclear weapons have sophisticated electronic locks. Any attempt to force the lock will render the weapon defused. So whatever is offered on the black market, experts say, is useless.

All right then, you're not going to buy one. Are you going to make one?

To do so, you will need to get dozens of kilos of highly enriched uranium from somewhere. A precarious business, because the world's uranium supply is closely monitored. If even a milligram cannot be accounted for, a thorough investigation follows. If you *do* manage to obtain uranium, it still has to be enriched. That is specialist work that requires a giant plant with equipment that you would need to buy or build yourself. Something like that won't easily go unnoticed.

Break into a nuclear plant then, to steal the enriched uranium? That won't be easy. Taking out the guards. Sabotaging cameras. Forcing locks. In the unlikely event you reach the nuclear reactor, you will need a special construction crane to take out the fuel rods – remotely, of course, because close physical proximity to spent fuel might be lethal. Even if you get away with it unseen, you won't have nearly enough material for a nuclear weapon. On to the next nuclear plant?

Oh, and if you prefer to go for the fresh fuel rods that haven't been in the reactor yet, you will still need further enrichment.

Perhaps you can seize nuclear waste in equally inventive ways and pick out the plutonium? Alas, whatever waste comes out of a nuclear plant is also far too diluted.

By way of illustration of the complexity, when North Korea conducted its first nuclear test in 2006, the shock wave was, in fact, fairly limited. Based on measurements, experts suspect that the test failed. And this was a prestige project on which the country's brightest scientists had been working since the 1980s.

For the (even) less talented, of course, there is the 'dirty bomb'. This involves combining conventional explosives with radioactive material, which you might steal from the radiology department of a hospital. It involves highly skilled and dangerous work. Why go to so much trouble? And if such a dirty bomb is used – which has never happened yet – there will surely be considerable damage within a radius of perhaps 100m (330ft), but any further beyond, the damage is mostly fear. That's why experts refer to dirty bombs not as weapons of mass destruction, but weapons of mass disruption.

Far too complicated, far too expensive: these have always been key reasons as to why there aren't many more nuclear weapons. A range of countries, including Brazil, Sweden, Taiwan and South Africa, have at one time considered a nuclear weapons programme, or even started one, but have since decided not to pursue it. When the Soviet Union broke up, Ukraine, Belarus and Kazakhstan suddenly found themselves with nuclear weapons. They quickly sold them to Moscow.

Nuclear weapons? *No, thanks!*

Not only is it striking that we're not facing a whole lot more nuclear weapons, as many had feared, perhaps more surprising is that the total number of nuclear warheads is declining.

In the 1980s, the world's stockpile of nuclear weapons peaked at 70,000. Thirty-five years later, one in five of those remain. That's still more than can be rationally justified, but peace activists hardly dared to dream of such a sharp decline. Yet this uplifting news has largely escaped public attention.

How does one get rid of nuclear weapons? They're simply taken apart. 'It's like any other kind of machine,' says Robert Rosner, former director of the Argonne National Laboratory, which grew out of Enrico Fermi's work on the first nuclear reactor. 'It's a case of taking it apart piece by piece.'[20]

As an aside, Rosner is not one to take nuclear weapons lightly. He is closely associated with the Bulletin of the Atomic Scientists, a nonprofit organisation that has been warning of nuclear war since 1945 and is the keeper of the Doomsday Clock, which it continues to adjust periodically, keeping it close to midnight.*

So, it's not terribly tricky to dismantle a nuclear weapon, Rosner said in an interview with *Inside Science*, the news service of the American Institute of Physics. The design is what he calls 'an open secret'. In a scenario in which Americans had to deal with bombs from North Korea, Rosner says, 'it wouldn't be much of a mystery to them'. Rosner is also convinced that it's 'very unlikely that it would blow up if a mistake was made in the process of disassembly.'

Well, that's a relief. Apparently, it's a lot easier to do away with nuclear weapons than to get them.

But what to do with the removed uranium and plutonium? One option is to mix it with other ingredients and turn it into fuel for a nuclear plant. Both the American and Russian governments decided that many thousands of nuclear warheads from the former Soviet Union would be blended down to provide electricity in the United States. At the height of the 20-year Megatons to Megawatts Program, whereby the US bought 500,000kg (1.1 million lb) of Russia's weapons-grade uranium for use in its nuclear power fleet, this provided 10 per cent of the nation's power production.

As journalist Gwyneth Cravens nicely put it in *Power to Save the World*, the bombs that were meant to blow up cities were used to light them instead.[21]

It's a pretty image: the world's most dreaded bombs, silently tucked away in a power plant which releases all that carefully

---

* The Doomsday Clock is a metaphor to help the public visualise how nigh the end of the world is, with midnight representing global catastrophe. The 1947 setting was at 7 minutes to midnight, and as of January 2024 is now at 90 seconds before midnight.

compressed force of destruction in a highly controlled manner to fulfill the role of a humble servant through our refrigerators, vacuum cleaners, toasters, electric blankets... It couldn't be further removed from their original purpose. Yet from the very beginning – even *before* the beginning – the prospect of the atomic bomb had everything to do with peace.

In World War I, Leó Szilárd was an army officer. While he was recovering in hospital after being felled by Spanish flu during the war, he learned that almost all his regiment comrades had died on the Italian front line. Since then, Szilárd dreamed of peace. How could war be avoided? It was a question that completely absorbed the freethinking Szilárd. His answer: war is avoided when you possess a weapon so powerful that no country dares to attack you.

Like his contemporaries, Szilárd suspected that somewhere in the atom, a terrifying power is hidden. Unravel the secret, turn it into a bomb and you bring every warmonger to his knees. With such weapons in existence, who would ever start a war?

During World War II, Szilárd conveyed this vision to his colleagues at the Manhattan Project. Their work was never intended to wage war, but to end it. In fact, Atoms for Peace, the famous slogan that President Eisenhower used to advertise nuclear power in the 1950s as a radical break with the past, had always been the motto.

The desire for a weapon powerful enough to prevent wars is an old one. For instance, the Swedish inventor of dynamite – one Alfred Nobel – believed that his mixture based on nitroglycerine, which he patented in 1867, would surely put an end to all armed conflict. He dreamed of producing 'a substance or a machine of such frightful efficacy for wholesale destruction that wars should thereby become altogether impossible'.[22]

Legend has it that the 'dynamite king' had regrets at the end of his life. A French newspaper is said to have mistakenly reported his death in 1888, when in reality his brother, a well-known businessman, had died: '*Le marchand de la mort est mort*' (The merchant of death is dead). Deeply disturbed by this painful headline, Nobel, as owner of nearly a hundred munitions and explosives factories, is said to have decided to donate his fortune to promote science and peace.

It's a nice story, but evidence is lacking. The newspaper bearing that headline has never been located, and Nobel was seemingly never publicly ashamed of his work.[23] In 1891, he told a pacifist baroness that his dynamite factories were more likely to end the war than the conferences she organised. 'On the day that two army corps can mutually annihilate each other in a second', Nobel said, 'all civilised nations will surely recoil with horror and disband their troops.'[24] The namesake of the Nobel Prize considered himself a pacifist, convinced that science could produce a weapon that would enforce peace.

Where some dreamed of such a weapon bringing peace, the same weapon caused nightmares for others. After all, the destructive power had to be immense. And so it was that Winston Churchill, Britain's future prime minister and a prolific writer, claimed that war had become 'the potential destroyer of the human race'.[25] Psychoanalyst Sigmund Freud argued that humanity had 'gained control over the forces of nature to such an extent that with their help they would have no difficulty in exterminating one another to the last man'.[26]

Both men wrote these words in the 1920s when the atomic bomb had not yet arrived. So, what were they referring to? Poison gas and aeroplanes. As political scientist John Mueller observes in his clever and refreshing book *Atomic Obsession*, the use of poison gas made people shudder. In World War I, gas could blind, suffocate and disfigure like no other weapon. And then there were the fighter planes. Many citizens had never seen even an ordinary plane before; now they came to scatter bombs over their city. The Germans bombed the British to break their morale and 'provide a basis for peace'.[27]

As Europe licked its war wounds, new theories emerged. Poison gas 'may well prove the salvation of civilisation', a leading British military analyst speculated in 1925.[28] Precisely because the stuff was so nasty, no army leader would want to use it ever again. With similar reasoning, chemist Fritz Haber had convinced the German military a decade earlier that the awful poison gas he was working on would shorten the war and thus save lives.

Air raids were equally pernicious, especially now that aircraft could be more heavily loaded and fly further. If a tonne of gas was scattered over Paris from 100 aircraft the city would be wiped out in an hour. In 1938, a military expert wrote:

> *The very magnitude of the disaster that is possible may prove to be a restraining influence. Because the* riposte *is certain, because it cannot be parried, a belligerent will think twice and again before he initiates a mode of warfare the final outcome of which is incalculable. The deterrent influence may, indeed, be greater than that. It may tend to prevent not only raids on cities but resort to war in any shape or form.*[29]

Soon, the experts must have felt rather disappointed.

But of course, the atomic bomb surpassed everything! This was, in the words of H.G. Wells, the 'ultimate explosive'.[30] Oh yes, the atomic bomb would put an end to all war, would it not? The bomb was given mythical status and so one day, it gained a capital letter: the bomb became the Bomb.

If you have one, you'll keep the enemy at bay. Leonid Kravchuk, Ukraine's first president, realised this too late. In the early 1990s, he handed over 1,700 nuclear weapons to Russia in exchange for security guarantees, but years later, in 2014, he had to watch as Vladimir Putin dispatched his army and annexed Crimea: a new low after years of threats against Ukraine. 'Had we kept the nuclear weapons', Kravchuk said, 'they would have treated us differently now.'[31] In February 2022, his country was invaded by Russia again.

Had Nobel, the dynamite king, been alive today, he might well have awarded the creators of the atomic bomb a Nobel Peace Prize.

In short, the whole point of the atomic bomb is not its use, but the threat of using it. Or so goes one theory in circles of foreign policy experts. But is it really just a threat? When a mugger on the street puts a gun to your head, is he merely *threatening*, or is he actually *using* his weapon?

Besides, not everyone has been scared off. Israel is widely believed to have had nuclear weapons since the late 1960s, but was attacked nonetheless in 1973 by a coalition of Arab countries, led by Egypt and Syria. The UK had nuclear weapons when Argentine military forces occupied the Falkland Islands in 1982. In 1999, a year after Pakistan tested its atomic bomb, war briefly broke out with India: the only time two countries with nuclear weapons have gone to war with each other. In 2006, Hezbollah militia in Lebanon attacked an Israeli border post. In 2023, Palestinian militant groups launched a surprise attack on Israel from the Gaza strip.

Deterrence does not always work, not everywhere, and not with everyone.

Moreover, the practical usefulness of nuclear weapons has proved to be quite limited. They were not used in the conflicts mentioned above, and neither were they used when the Americans went to war with the North Koreans or the North Vietnamese. Nor did China use them against Vietnam. France didn't use them when Algeria fought for independence. The UK didn't use them when it fought Egypt over access to the Suez Canal. The Soviet Union didn't use them against Afghanistan's Muslim fundamentalists.

That's why Mueller concludes in his book that nuclear weapons are simply not very useful.

Just as peace activists questioned the strategy of nuclear deterrence, experts questioned the military utility of nuclear weapons. This began as soon as the damage to Hiroshima and

Nagasaki was first recorded. Yes, many houses had burned down, but then again, these were made of wood. Buildings made of steel or concrete were still mostly standing. Yes, the railways, the streets, the underground water supply were all damaged, to be sure, but not devastated. Had the US Air Force carried out some 300 conventional aerial bombings over these two cities, there would have been at least the same amount of damage.

By the end of World War II, the Americans had become quite adept at aerial bombardment. This was partly thanks to their state-of-the-art bomber, the infamous B-29, whose development and production was far more expensive than the entire Manhattan Project. In Japan, American soldiers began by dropping explosives that detonated in the air, blowing away the roof tiles of houses. Then a rain of incendiary bombs, filled with napalm or white phosphorus, for example, set off the flames. A firestorm ensued, burning everything to the ground and destroying the water and gas pipes.

Tokyo was a regular target of such bombings. But things were never as bad as on the night of 9–10 March 1945. Nearly 300 bombers scattered incendiary bombs over the densely populated centre for hours. Everywhere, Tokyo residents met a wall of fire that must have measured around 1,000 degrees Celsius. The heat caused people and buildings to burst into flames. Those who jumped into rivers and canals were boiled alive. The stench of burning flesh crept up into the air, so high, in fact, that the bomber crew, high above the city, reached for their oxygen masks to avoid vomiting.

One hundred thousand men, women and children died in Tokyo within six hours. Today, that's still a macabre world record. Yet the massacre hardly raised any criticism from the American people or European politicians.

Dozens of Japanese cities were levelled that way. When the atomic bomb was ready for use in the summer of 1945 and the Americans were looking for a city that was still standing so they could properly measure the damage, few options were left.

Compared to the destruction caused by regular air raids, the effects of the atomic bomb, were not that special. Would it really have been less bad if Hiroshima and Nagasaki had been hit by hundreds of regular bombs?

A big difference is that an atomic bomb comes with radiation that might still cause diseases after many years. There is, as we have seen, proof of long-term damage caused by radiation, although not nearly as much as first expected. Since the 1950s, the well-being of nearly 100,000 survivors in Hiroshima and Nagasaki has been closely monitored. By the turn of the century, a portion of them had died from a variety of diseases and ailments that can affect us all as we age. The group included no more than 1,000 excess cases of cancer.[32]

Back to 1945. In its 13 August edition, *Life* magazine published six gruesome photos showing the burnt remains of several Japanese victims who had remained in their hiding place. The caption spoke of 'easily the most cruel, the most terrifying weapon ever developed'.[33] The new weapon in question was, brace yourself: the flamethrower. Indeed, as the atomic bomb fell moments after America's most renowned weekly went to print, the photo editor displayed a sad truth: weapons and war are, by definition, hideous.

Little wonder, then, that the Japanese didn't surrender immediately after the first atomic bomb was detonated. For them, Hiroshima was yet another city reduced to rubble and ashes. They didn't perceive the bombing on 6 August 1945 as a radical break with the past at all.

Three days later, after US pilots scattered millions of pamphlets over dozens of cities urging surrender, the bombing of Nagasaki took place. Still the Japanese didn't surrender. At a top-level military meeting, the main concern was the Russians, who had unexpectedly invaded the country.

Almost a week passed. The Americans wondered why the Japanese had not been impressed by their atomic bombs. Waveringly, they were preparing a third when Emperor Hirohito realised that he could end the war and save his own skin. Upon Japan's surrender, he spoke of 'a new and most cruel bomb, the power of which to damage is indeed incalculable'.[34] The peace negotiations stipulated that Hirohito would be allowed to remain as emperor and would not be charged with war crimes.

As some historians have argued, both the victor and the loser now had good reason to emphasise the extraordinary destructive power of the atomic bomb: the Americans because it enhanced their international prestige, the Japanese because it gave them an excuse for defeat. They had lost not because of failed leadership or lack of courage, but because of a chilling feat of engineering conceived by the heartless enemy.

To this day, there are vested interests in keeping the myth of nuclear weapons alive. American officials cultivated a narrative that the Japanese people would have fought to the death, even though nobody protested over the surrender. In fact, when literary scholar Masao Miyoshi reflected on his training as an air force cadet to become a kamikaze pilot in 1945, he said that 'all the cadets were thrilled not to have to fight' when they heard that the emperor had finally surrendered. 'They threw their hats in the air and cheered.'[35]

It looks very much as if the atomic bomb didn't so much mark the end of World War II as the start of the Cold War.

During the decades of geopolitical tension between West and East, the Americans appear to have been fooled by the Russians. Nikita Khrushchev, First Secretary of the Communist Party from 1953 to 1964, boasted that Russia was producing nuclear weapons as if they were sausages. During a reception at an embassy, he snapped at Western diplomats: 'We will bury you!'[36] It was all grandiose talk; there is no evidence to suggest any serious plans for the Soviets to be the first to fire the weapons.

The American government would never readily admit to being misled by the Russians for decades, potentially wasting untold amounts of money and talent on developing, maintaining and stockpiling weapons that, practically speaking, were not only useless, but also unnecessary politically.

Mueller's conclusion is sobering: both proponents and opponents have grossly overestimated the power of nuclear weapons to change the world. No, nuclear weapons did not lead to a war that destroyed everything and everyone, and no, neither did they lead to a peace that reconciled everything and everyone. Overall, the impact of nuclear weapons on world history has been quite limited.

But please don't tell anyone! Perhaps it's better if they keep us terrified.

The World Set Free involves a plot twist that once again shows foresight. Where Wells has Holsten, the inventor of the atomic bomb, strolling through London in the first pages of the book, Leó Szilárd was strolling through the same city when he had his own lightbulb moment. However, while Szilárd was full of hope for world peace, Holsten was full of remorse:

> *He was oppressed, he was indeed scared, by his sense of the immense consequences of his discovery. He had a vague idea that night that he ought not to publish his results, that they were premature, that some secret association of wise men should take care of his work and hand it on from generation to generation until the world was riper for its practical implication. He felt that nobody in all the thousands of people he passed had really awakened to the fact of change; they trusted the world for what it was, not to alter too rapidly, to respect their trusts, their assurances, their habits, their little accustomed traffics and hard-won positions.*[37]

Szilárd, too, was beset by remorse after a while. When the atomic bomb was nearing completion, he no longer seemed keen to use it. Perhaps, he suggested, the Americans should issue a warning first? Maybe they should just organise a demonstration and invite the Japanese? In a letter to the president, signed by dozens of Manhattan Project employees, Szilárd asked for wisdom to be applied.

When he learned of the Hiroshima bomb, Szilárd described it as 'one of the greatest blunders of history'.[38] He left nuclear physics and became a biologist: he no longer wished to be concerned with death, but with life.

Where Szilárd's hope turned into remorse, Holsten's remorse, in the book, turned into hope – and not just because, as Wells writes, he built an engine which in 1953 'brought induced radio-activity into the sphere of industrial production, and its first general use was to replace the steam-engine in electrical generation stations',[39] opening a way for mankind 'to worlds of limitless power' and crossing a 'new boundary in the march of human progress'.[40]

Above all, Wells describes how virtuous technocrats learned their lesson. A diplomat, obsessed with the idea of ending war, organised a conference for world leaders to 'save humanity'.[41] Together, they decided to form a kind of world government, which would build a harmonious global society. Thanks in part to international cooperation and the surveillance of Carolinum – 'we have to get every atom of Carolinum and all the plant for making it, into our control', said one attendee[42] – the foundations were laid for a new chapter in humanity: a period of peace, freedom and progress.

The world set free.

On 24 October 1945 the United Nations was officially established. The opening sentence of the Preamble to the founding treaty revealed the founders' ultimate goal: 'to save succeeding generations from the scourge of war'.[43] Sixty years later, the International Atomic Energy Agency, the UN organisation that oversees the peaceful use of uranium, was awarded the Nobel Peace Prize.

CHAPTER EIGHT

Perverse Incentives

Did the nuclear industry start the decline of nuclear energy?

'Hello, this is your captain speaking: there is absolutely no cause for alarm.'

– Monty Python, in *How to Irritate People*, 1968

Why does it seem as if nuclear power has become such a failure? A common answer goes like this.

Action groups had been demanding strict safety rules for years when, in 1979, a serious accident occurred at the Three Mile Island nuclear plant near Harrisburg. Immediately, the advance of nuclear energy stalled, starting with the United States, the global industry leader. When disaster hit Chernobyl in 1986, it was another blow, this time mainly in Europe. And when in 2011 a tsunami sparked a meltdown in three nuclear reactors in ultra-modern Japan, resulting in an extremely expensive clean-up operation, many took this to be the death knell. Building a nuclear plant that can meet safety standards now costs so much, and takes so long, that nuclear power is finished. Accidents have crippled the industry.

This popular explanation excels in simplicity, but it's not the whole story. The decline began much earlier. In fact, as is often the case in the incredible story of nuclear power, it's exactly the other way around. What crippled the industry was not the idea that an accident was quite possible, but rather the idea that an accident was utterly *impossible*.

To understand this, we need to go back to the 1950s, when the most brilliant scientists and engineers were building a completely new industry. They were aided by the world's most powerful government, which touted nuclear power's endless benefits everywhere. In America, nuclear power was all the rage.

There was no shortage of self-confidence. The chair of the US Atomic Energy Commission (AEC), which fostered and controlled nuclear science and technology, claimed that energy would become 'too cheap to meter'.[1] Nowadays, with nuclear plants costing billions more than budgeted, this statement is sometimes smugly referenced by nuclear's naysayers: *haha, such optimism...* But in 1954, the AEC's prediction was not so outlandish. Rarely would an industry grow so large so quickly. Within 15 years, nuclear power was already just about competitive with coal, which had long been widely used, as well as oil, that other competing fuel in the fast-growing electricity market.

Nuclear power was becoming more and more interesting, if not to make weapons, then for its potential to make the West less dependent on Middle Eastern oil exporters, or else to put pressure on coal miners who kept striking for higher wages. A sense of superiority settled in.

If something went wrong in a nuclear plant in these early years, it was hushed up. If a nuclear physicist made a critical comment about what on earth to do with the waste, or about the dark link with weapons production, or about the risks when radioactivity is released, soothing words or hard-nosed denials were quickly heard. There was nothing to worry about, everything was under control.

By the 1970s, an accident was virtually unthinkable – at least to the 'experts'. Two examples illustrate this haughty attitude. The AEC director of reactor safety, speaking to a news agency in 1974, said: 'There will never be a major accident in a nuclear power plant.'[2] An industry representative, speaking to a journalist, in 1979, said: 'Frankly, I don't believe a serious accident could ever happen.'[3] These remarks were exposed as arrogant lies made by a self-righteous technical elite.

A striking example came at a press conference after the events at the Three Mile Island nuclear plant. A vice president of the power company that owned the plant was clearly bothered by journalists who simply wanted to know what had happened and what was being done. At one point, he said: 'I don't know why we have to tell you each and every thing we do.'[4]

So how did the world's leading nuclear industry – that of the United States – react when a rather odd protocol was proposed in the first half of the 1970s that would lead to far-reaching safety requirements for all nuclear plants? There was some grumbling, but, full of complacency, they accepted it. They thought nothing could go wrong.

But things did go wrong, because the protocol (more on this later) made it difficult to supply nuclear energy more cheaply than fossil-based energy. While the coal industry did not have nearly as many regulations around safety and the environment, and was able to work more efficiently and cheaply thanks to innovation, nuclear power saw an explosion in regulation. The cost of building a nuclear plant rose dramatically.

To better understand this striking turn of events, we'll need to travel through time. In this chapter, we'll go back to the years when Americans were concerned about the fallout from atomic tests, and the nuclear industry suspected that a lower radiation limit could alleviate concerns. Moving back even further along the timeline of defining moments in nuclear history, we'll arrive where it all began, with a biologist and his collection of fruit flies. And then, leaping forward, we'll reach the modern day, a time during which authorities are spending big money to reduce the risks of nuclear power. We'll stick mostly to the United States, since this is where both the rise and fall of nuclear technology began.

But first: what safety protocol are we talking about?

To protect staff working at a nuclear plant and the people living nearby, agreements are made on the maximum permissible additional radiation dose. Obviously, this limit is set on the basis of scientific evidence regarding potential damage to health. Then the legislator sets the limit somewhat lower, simply as a precaution, followed by independent, close monitoring to ensure that the owner of a nuclear plant doesn't exceed the agreed limit.

But that's not how it works in reality. This may be how things operate in every other industry, but not in nuclear power.

Rather, the deal is that the owners of a nuclear plant must do everything in their power to ensure the radiation isn't simply below the limit, but as far below as possible. This protocol is called ALARA, or 'as low as reasonably achievable'. It is the gold standard for radiation protection.

What does 'reasonably achievable' mean in practice? It means that opportunities to increase safety must always be pursued, even when the need is questionable at best. If an extra set of pumps and valves can be added, then there will be an extra set of pumps and valves added. If a nuclear plant is profitable, the regulator may come knocking on the door: could that concrete wall be made a bit thicker please, because apparently there's the money to do so.

The excessive regulation of nuclear plants is in stark contrast to the relaxed attitude towards coal plants. While the regulator is reluctant to issue licences for nuclear plants that do not emit air pollutants or greenhouse gases, power plants burning fossil fuels that emit particulate matter, nitrogen and sulphur and are said to bring premature death to millions of people every year have it much easier.

But E.F. Schumacher, the German environmental guru who worked in the British coal industry, liked to stress that we shouldn't overestimate the problem of air-polluting coal plants. In *Small Is Beautiful*, he questioned why anyone would insist on clean air since 'the air is laden with radioactive particles'.[5]

To find out how regulations became so strict, let's go back to the 1950s once again. The Americans were on the winning side in World War II, but they were afraid. Nuclear weapons – *their* nuclear weapons – were making the world unsafe. While no one dared to speak about the horrific consequences of a nuclear attack, concern shifted to the mysterious radioactive particles that drift and descend with the wind after an atomic test. Might that radiation cause cancer? Might it cause birth defects in children?

Evidence for such effects is lacking, but neither is there evidence that exposure to a small amount of radiation causes no harm at all. Responding to public unrest, the experts sharply reduced the maximum permissible dose limit in a matter of years, from 250 mSv per year in the 1930s to 150 mSv in 1949, and then to 50 mSv in 1956. (It's now 20 mSv.)

These limits apply to professionals working in nuclear plants, weapons laboratories and hospitals, but the public must of course be protected, too. Civilians, it is decided, should not be exposed to more than a tenth of the radiation acceptable in the professions. And what about children? More new standards! For minors, the exposure should be even lower.

Following the international ban on atmospheric nuclear testing in 1963, the debate on fallout subsided. But not for long. Ten years later, another oil crisis kicked in. Sheikhs in the Middle East increased the price of oil four-fold. In a 1973 address to the people, President Richard Nixon told of how he was planning to deal with what he called 'the energy emergency'.[6] The thermostat must be turned down a bit. People should be discouraged from flying. Available oil would be rationed. And he argued that the time to license a nuclear plant and complete construction should be reduced from ten to six years.

Nixon wanted his country to produce all its own energy before 1980, by increasing nuclear power's share in the mix. But, he warned the 400 journalists at a packed press conference, following his televised address: 'Don't write an editorial on this. You're really going to catch it from your readers if you do, because it scares people. Nuclear power. They think of the bomb. They think of the possibility that one of them is going to blow up.'[7]

Indeed, nuclear power was viewed with suspicion. Nuclear plants were said to be a radiation source. Local residents were alleged to suffer from all kinds of unexplained health issues. Activists hung posters in the areas where nuclear plants were in operation or being proposed, warning of exposure to radioactivity. Some argued that a wide radius around nuclear plants should be declared a 'danger zone'.[8]

The nuclear clique was increasingly seen as corrupted. In the US, nuclear power was promoted by the same organisation (AEC) that regulated it. Such a double act was no longer acceptable. The AEC was split into two operations: the Nuclear Regulatory Commission, tasked with protecting health and safety, and a separate agency that soon became part of the US Department of Energy and took charge of research and development.[9]

The new regime didn't go unnoticed. In the early 1970s, there were about 100 guidelines and standards for nuclear plant design and construction. By the end of the decade, there were more than 2,000, and a new one was added every day.

'Our industry is the most heavily regulated industry in the world', observed Richard 'Rip' Anderson, who worked in a nuclear energy research laboratory.[10] In a 2007 book by Gwyneth Cravens, *Power to Save the World*, he's quoted as saying: 'We have two NRC inspectors here all the time. It's like having a highway patrolman in your car to make sure you don't speed.'

In the 1970s, safety was becoming the focus of the industry. Once ALARA was introduced, *everything* needed to be done to ensure maximum radiation protection and accident prevention. The rules and standards applied not only to new reactors, but also to those already under construction. Entire sections were demolished and rebuilt. Because of the rules, the amount of steel needed to build a nuclear plant doubled. The amount of concrete tripled. A simple part like a valve or the tools to maintain it suddenly cost many times more in a nuclear plant than in a coal facility. The maintenance of pumps and heat exchangers was regulated to such an extent that a simple routine job turned into a time-consuming project.

Within a generation, everything had changed. The promise of becoming the cheapest energy source ever was unfulfilled. Facilities took more and more time and more and more money to build. It became perfectly normal for a nuclear plant to be finished much later and cost much more than planned. As rules were constantly being tightened, ordering and building several identical reactors at once became impossible. Safety rules prevented standardisation.

After 1975, decades went by with no permit being issued in the US to build a new nuclear plant. Gradually, dozens of construction projects were halted and demolished. Hundreds of plans were discarded. New plans failed to materialise.

The industry, which still had not experienced a serious accident at a nuclear plant, adopted ALARA in an attempt to reassure the public. But it didn't work. The public did not want to be soothed. And after all, the obsession with safety was not unreasonable. If the industry itself treated very low doses of radiation with the utmost care, it must be very dangerous, right?

Indeed, the flood of new safety requirements created the impression that something was always wrong, and that public health was at stake. It led to even more suspicion. Weren't older nuclear plants unsafe by the new standards? How dare the industry be so irresponsible for so long? And why would these new requirements be any safer? Why should we believe such liars?

But all this time, of course, it wasn't radiation limits or safety standards that needed adjusting. It was fear.

Looking back, we can see that it wasn't opposition to the plants, but opposition to the weapons that became the biggest obstacle to the development of nuclear power. When industry growth slowed down in the West, the movement that would come to oppose nuclear plants was still in its infancy. At the beginning of 1975, the construction site at Wyhl am Kaiserstuhl in Germany (see Chapter 3) had yet to be occupied and Greenpeace didn't even have a US branch.

The anti-nuclear movement is still pretty busy, to this day: protesting against plans for new nuclear plants; delaying procedures for permits; signing petitions for closure of those 'satanic mills'. Activists pat themselves on the back for their successes. But in fact, the industry had already dug its own grave. The activists only had to push it over the edge.

Let's jump back further in history, to the time when physicists were enthralled by radiation. In turn, biologists were captivated by another invisible phenomenon: genes. One of the bright minds keen to learn all about this hereditary material was a young, ambitious student of humble origins from New York City: Hermann Muller. He would be at the forefront of the ALARA protocol.

In 1915, while his contemporaries were being drafted to fight in Europe, Muller had other things on his mind. He was studying fruit flies in a laboratory. Muller wanted to know whether genetic changes could be transmitted to offspring. Fruit flies were ideal for the work: they're fully grown after only 10 days, the females produce hundreds of eggs, their chromosomes are easily visible under a microscope and they need little more than some old banana skins to eat. Also, you never hear them complain.

Unfortunately, it's not exactly exciting studying *Drosophila melanogaster* one by one for years on end, observing very few mutations.

However, that all changed when Muller borrowed an X-ray machine to see if the rays discovered by Wilhelm Röntgen would have an effect on his test subjects. He was startled by the results. The radiation made the flies instantly sterile. When Muller adjusted the radiation dose to just below the level at which they became infertile, he saw to his dismay that there were 150 times more mutations than normal. If he halved the dose, the number of mutations halved too. If he halved the dose again, he saw another halving of mutations. He had such a huge number of fruit flies and he started at such an overwhelming dose of radiation that it went on and on. It was 1927, and Hermann Muller had finally found something: a link between radiation and genetic mutations!

Muller made his discovery at a time when diseases were beginning to appear in those health professionals who worked with X-ray machines or radium. Sooner or later, they perished from cancer, that debilitating disease so surrounded by fear and taboo. Muller's research appeared in *Science* as a discussion paper, with no data, no references and no explanation of methodology.[11] It caused a stir in medical circles. Could radiation lead to cancer? Could the damage to health from radiation also live on in offspring?

That same year, the US economy fell into recession, leading to the Great Depression. Muller noted with disappointment that less money was available for his laboratory work. The promised scientific support for his article in *Science* never materialised.

Muller threw himself into eugenics, the study of genetic makeup that had fascinated him since his student days. He regularly spoke at international conferences on how eugenics could perfect the human race. He launched a manifesto advocating genetic improvement of the world's population. His son's middle name was a tribute to this field: Eugene.

Muller also developed a sympathy for communism. In 1933, he left for the Soviet Union, where he wrote a book on eugenics, *Out of the Night*. He had a translation delivered to Joseph Stalin and urged the development of a programme of artificial insemination with sperm of superior genetic quality that would be supportive of the socialist utopia. The Soviet leader was not a fan; he considered this whole eugenics thing to be merely bourgeois.

Concerned for his own safety, Muller packed his bags and got out of the country. He travelled across Europe with 250 species of fruit flies. When World War II started, he returned to the United States, disillusioned and penniless. Fortunately, he was presented with an opportunity, which he grabbed with both hands. He was invited to do consultancy work for the government. Only later did Muller realise that his work must have been linked to the development of the atomic bomb.

In fact, it was 'the Bomb' that would put his work with fruit flies back into the limelight. On 12 December 1946, Muller received the Nobel Prize in Physiology or Medicine for discovering that radiation can cause genetic mutations. The Nobel committee's decision to reward someone for a nearly 20-year-old study on fruit flies appeared to some to be a political comment: the atomic bomb should be rejected.

In his Nobel lecture, Muller explained how he had clearly shown that 'the frequency of the gene mutations is directly and simply proportional to the dose of irradiation applied'.[12] Whether it was alpha, beta or gamma radiation, and whether one accumulated the dose all at once or built it up over a long period of time, made no difference. The facts, he said, leave 'no escape from the conclusion that there is no threshold dose'.

No threshold dose. That meant even a tiny amount of radiation has adverse effects. This observation led to a hypothesis that would come to be known as linear no-threshold, or LNT for short. It was based on the idea that more exposure to radiation leads to proportionately more damage to health and that there's no safe lower limit.

With a Nobel Prize in his pocket, Muller finally became a man who was listened to. At the United Nations, he advocated for a ban on atomic testing and signed a manifesto warning about the deadly radioactive particles released into the air when a nuclear weapon goes off. Radiation exposure among local residents after an atomic test was rarely higher than natural background radiation, but that nuance escaped the public. When even a small amount of radiation leads to cancer, according to the prevailing view, every atomic test is one too many.

Muller's influence reached its peak in the mid-1950s, when he participated in a major study on the biological effects of radiation as part of a National Academy of Sciences subcommittee. This ought to be clarified once and for all, according to some, because of all those atomic tests and the aggressive state propaganda for nuclear power. Muller didn't like these peaceful applications either; he ended his Nobel lecture by warning against the increasing use of *any form* of nuclear energy. He even forbid his children and grandchildren from having X-rays taken at the dentist.[13]

Now Muller and other scientists were taking stock and making recommendations to the government. His ideas featured prominently when the study was released on 12 June 1956 before a modified version was eventually published in the June edition of *Science*.[14]

The report stated that the radiation released from nuclear fission led to changes that are 'often harmful' and can be passed on to future generations. There was 'no minimum amount of radiation' before mutations occur. The damage was cumulative. Much of the harm might go unnoticed at first, but it was there and would live on in offspring. With exposure to more and more radiation, the rising death rate and falling birth rate would result in the population eventually declining. Therefore, we should limit any radiation exposure as much as possible because, 'from the point of view of genetics, [it is] all bad'.

The enthusiasm for President Eisenhower's planned construction of nuclear plants must be tempered, the subcommittee concluded. All forms of radiation should be kept 'as low as possible'. Anything that added radiation to 'the naturally occurring background rate' would cause mutations and was 'genetically harmful'.

Waste in particular would become a major source of concern. If so many nuclear plants were really going to be built, we would be surrounded by radioactive waste in 'almost unthinkable quantities', which must be kept separate from humans and the environment for centuries to come. 'When a world-wide atomic power industry becomes fully developed, its accumulated waste products might represent more radiation than would be released in an atomic war.'

The report's conclusions had long since moved beyond atomic testing. A 'safe and rational growth' of the nuclear industry meant, the authors argued, that you couldn't just build a nuclear plant anywhere, because its presence 'will be felt, in various ways, over a wide region'. Accidents were 'inevitable' and everything must be done to ensure they did not become 'catastrophes'.

For the record, the initiator and backer of the study, which assembled the team of experts, supervised the research process and co-wrote the final text, was the Rockefeller Foundation, the philanthropic institution of an oil tycoon with most of his family fortune tied up in shares in the fossil fuel industry.[15] The subcommittee was chaired not by a distinguished peer, but by a mathematician who worked as a director for the Rockefeller Foundation.[16] The research appeared under the banner of the US National Academy of Sciences (NAS), whose top executive also served as president of the Rockefeller Institute for Medical Research.[17]

The *New York Times* considered it important news. As many as six articles, including three full pages of verbatim text from the report, appeared in its 13 June edition. The headline at the top of the front page read: 'Scientists Term Radiation a Peril to Future of Man'.[18] This is how the article began: 'A committee of outstanding scientists reported today that atomic radiation, no matter how small the dose, harms not only the person receiving it but also all his descendants.' With the report getting traction, a solid foundation was established for the belief that radiation was a unique hazard that had no safe dose. The LNT hypothesis was born.

By the way, the publisher of the *New York Times* was affiliated as a trustee of the Rockefeller Foundation.[19]

A strange assumption is at the heart of LNT. According to the theory, it doesn't matter whether you contract radiation all at once, such as during a bomb blast, or build it up slowly over decades because you happen to live in the mountains or work as a flight attendant. But that's odd. Our bodies have found all sorts of ways to repair cell damage, especially when it occurs in a measured way.

LNT also leads to bizarre arithmetic. Some experts came up with entertaining analogies to show just how bizarre. If a person dies after taking 100 aspirin in one go, then according to LNT arithmetic, there would also be one death if 100 people took one aspirin each.[20] Or, if 100 people step off a footstool 30cm (12in) high, you would get the same effect as if one person jumped from an apartment building 30m (100ft) high.[21]

If all this sounds puzzling, it's because it is. LNT is truly strange.

Despite protests from scientists, the Rockefeller committee's recommendation for LNT was loosely transferred from fruit flies to the human species. The unwieldy array of committees, agencies and governing bodies around radiation protection were under public pressure to adopt LNT, although they remained on the fence about it.

UNSCEAR stated that it was difficult to find 'reliable information about the correlation between small doses and their effects either in individuals or in large populations'. It explained that 'linearity has been assumed primarily for purposes of simplicity' and stressed that it was not clear whether 'there may or may not be a threshold dose'.[22]

The Joint Committee on Atomic Energy (JCAE), a US government legislative organisation, emphasised that no harm had been observed at low doses. It 'strongly' questioned the assumptions and therefore could not determine whether there is a threshold dose.[23]

The International Commission on Radiological Protection (ICRP) believed that the assumptions 'may be incorrect', but concluded that 'they are unlikely to lead to the underestimation of risks'.[24]

The National Council on Radiation Protection (NCRP), in accepting LNT, stated that the adjustments to the permissible dose limit were 'not the result of positive evidence of damage'. Rather, they were based, the organisation continued cryptically, 'on the desire to bring the MPD (maximum permissible dose) into accord with the trends of scientific opinion'.[25] What might 'scientific opinion' be? Did it rely on facts or on views? And in whom exactly did this 'desire' reside?

In the following years, the tepid acceptance of the LNT hypothesis would lead to extreme measures that had nothing to do with safety and health, but everything to do with the need to protect and regulate. LNT was at the root of ALARA, the requirement that a nuclear plant owner must do everything possible to minimise radiation exposure. That race to zero is the practical translation of what Muller established as an inescapable conclusion: 'no threshold dose'.

Today, UNSCEAR still accepts LNT but acknowledged in a 2015 report that 'scientific results obtained during the past one or two decades challenge the LNT hypothesis'. The authors add that 'the Scientific Committee does not recommend multiplying very low doses by large numbers of individuals to estimate numbers of radiation-induced health effects within a population exposed to incremental doses at levels equivalent to or lower than normal natural background levels'.[26]

A former secretary of the ICRP has called LNT 'a deeply immoral use of our scientific heritage'.[27]

When a tsunami led to an accident that made Fukushima a symbol of nuclear catastrophe, just as the atomic bomb did to Hiroshima, the future of nuclear power was immediately at stake. To some, the waiting had been rewarded. *You see? We told you, it's not safe.*

Listen to the members of the Bulletin of the Atomic Scientists. This organisation, set up by a remorseful member of the Manhattan Project, is an established voice in all things nuclear. What did the Bulletin think of events in Japan? With approval, it quoted a Japanese professor advocating a ban on nuclear power[28] and argued that the radiation released at Fukushima 'would destroy the world environment and our civilisation'.[29] Indeed, nuclear power stations, which have 'catastrophic potential', the Bulletin writes elsewhere, 'are just too dangerous to exist'.[30]

Interestingly, we never hear anything like this after an accident involving other energy sources. A coal mine collapses. A gas pipeline bursts. An oil train derails. A dam breaks. A wind turbine catches fire during maintenance. A solar panel blows off the roof in a storm. These things happen and people die, but no one demands that an entire industry be shut down. Maybe it's because people have been dying in floods, fires and storms for thousands of years. But when radiation is released from a nuclear plant? *Oh yes*, we think, *this is a unique hazard...*

Energy production, by any means, is not a flawless activity. Mining for the resources, transporting the materials, building and maintaining the installations, processing the waste: all sorts of things can go wrong at any time, whether you have a gas power plant or a wind turbine. How many fatalities does each energy source cause throughout its life cycle, per unit of electricity generated?

Researchers have started tallying. In one study, published in *The Lancet* in 2007, two conclusions emerged:[31]

1. *For billions of people, access to modern, reliable energy brings many benefits that save lives. Indoor air pollution is reduced, and medicines and food can be stored better.*
2. *By far the safest way to provide that energy is with nuclear plants.*

It seems the authors didn't like the last bit. They stressed that nuclear power was 'controversial' and would not be available in time 'to help the crucial near-term reduction in greenhouse-gas emissions'. Duly noted.

The Lancet study did not include solar panels and wind turbines. Other scientists did include them for an investigation of accidents occurring at all low-carbon sources, published in 2016.[32] Their research brought them to three conclusions:

1. *By far the most deaths, a whopping 97 per cent, came from he production of one energy source: hydropower.*
2. *Per unit of electricity produced, the most accidents as well as the most deaths came from the production of wind power.*
3. *The risk of an accident or fatality was found to be lowest with the production of nuclear power.*

The lead author must have felt uncomfortable. In a letter to *Science*, he argued that the debate over nuclear power is 'a damaging distraction' in the fight against climate change.[33]

Nuclear power is the safest of all energy sources: that's by no means the impression we get when we read a newspaper or watch the news. It's not even the impression given by nuclear experts. Let's look at an UNSCEAR report on the radiation levels to which workers and residents of Fukushima were exposed. The authors talk about 'the release, over a prolonged period, of very large amounts of radioactive material into the environment'.[34] That doesn't bode well.

Or take this quote from an interview between *Meltdown* author Yoichi Funabashi and Charles 'Chuck' Casto, head of the US Nuclear Regulatory Commission's support team in Japan, which the commission was watching closely because Fukushima's nuclear plant was American-made: 'In the history of mankind, Fukushima was a battlefield where a fight was fought against physics and nature. Just one step short of war, but in a certain sense it was a harsher ordeal than war. In war, you have the choice of surrendering. But such a luxury wasn't allowed in Fukushima. They had to fight nature and physics to the bitter end.'[35]

Wait, what? Why? Did nobody tell Casto and the authors of the UNSCEAR report that they were talking about an industrial accident in which nobody was killed and in which no noticeable health damage from radiation is likely?

It just doesn't seem to be an option to argue that the radiation released in an accident has very little impact on people and the environment. At companies selling nuclear power or nuclear fuel, communications departments either don't exist or simply don't see it as their job to correct errors in the media. No one in the industry openly acknowledges that a nuclear plant accident is always a possibility, now and in the future, and that the consequences are, well, not the end of the world.

The facts about the safety of nuclear power contrast sharply with our ideas about its danger. That's not only a maddening observation. It raises questions about the nuclear industry that have to be asked. Why does even an accident without a single radiation death have to be presented as an exceptional disaster? Why does the industry itself hold on to the myth that every bit of radiation is a hazard? Indeed, why does the language of the nuclear industry itself resemble that of the anti-nuclear movement?

These are all relevant questions, but the answers are in the realm of guesswork, and they come in the form of new questions.

Is the need to protect motivated by guilt over the atomic bomb, leading to a deep desire to come to terms with the violent origins of the technology? Is that why no one must ever again be harmed by nuclear energy?

Was Robert Oppenheimer referring to this when, after the detonation of the very first atomic bomb in the desolate American desert, he observed that he and his colleagues 'have known sin'?[36]

Does it perhaps confirm a need for recognition that, as in the days of Disney's 1957 TV show *Disneyland*, in the episode about 'our friend the atom', nuclear engineers are uniquely capable of navigating between life and death?

Or perhaps we need *more* nuclear accidents? After all, it may be precisely because they are so rare, that these accidents have acquired a mythical status, just like atomic bombs, which we don't see every day either.

One explanation for the industry's emphasis on safety comes from Bret Kugelmass, a successful tech entrepreneur who became fascinated by nuclear power and who hosts the Titans of Nuclear podcast. According to Kugelmass, it's not a matter of psychology, but of money.[37]

A meltdown in a typical nuclear plant poses no danger whatsoever to local residents, says Kugelmass. However, the industry has no interest in making this world news, he argues, for a simple reason: there's a lot of money to be made from the idea that any nuclear plant poses a unique health hazard.

When construction of new reactors came to a halt in the 1970s and 1980s, suppliers did not go bankrupt. The source of revenue didn't dry up, but merely shifted. With ever more safety requirements, there was always work to be done to existing reactors. They were in need of constant modification.

'In need of' became a flexible concept. After the disaster at Chernobyl, all US nuclear reactors had to be adapted to new safety regulations, even though they had a completely different design to the Russian reactors. The retrofits didn't make US nuclear plants any safer. But, says Kugelmass, a company like Westinghouse, which no longer built new plants, could send a $400 million (£320 million) bill for the upgrade. Per plant. So why complain?

If politicians decided to close a nuclear plant early, decommissioning followed. Such work must of course be done with extreme care, which takes years, even decades, especially at a plant where something has gone wrong. The reactor that partially melted down during the 1979 accident at Three Mile Island is not expected to be decommissioned until 2037. The cost is estimated at $1.2 billion (just under £1 billion). *Ker-ching!*

An accident, says Kugelmass, is a goldmine for everyone in the nuclear industry except the plant owner. He calls Fukushima, with its multi-year, multi-billion-dollar clean-up operation, the 'biggest payday yet'.

Others follow similar reasoning. In his book *Why Nuclear Power Has Been a Flop*, Jack Devanney mercilessly criticises our slavish acceptance of the exaggerated risks of radiation. As justification for continuing to pump hundreds of millions of dollars into it every year, Devanney argues, the industry must go along with the fantasy that even extremely low levels of radiation are harmful and with the demand that these levels must be constantly brought down further.

It must be unfortunate for these companies that accidents are rare. They may even dread the idea of it becoming widely known that there were no deaths at Fukushima due to radiation. That may explain why industry representatives keep quiet and don't challenge those who spread unfounded fear about nuclear power. That's why they willingly go along with the beliefs of Hermann Muller and others on LNT.

That, Kugelmass concludes, has been the revenue model since ALARA. Nuclear companies aren't so much selling electricity as protection. And who's paying? All of us. The costs are covered through our taxes and energy bills.

Because of climate change, today is the first time in history that nuclear energy is said to be a necessity. Will the industry suddenly make a real effort to debunk the myths and tell the truth? What professional radiation expert will say out loud that the fear is grossly exaggerated … and then look for other work? Which regulator or politician will dare to make the case for relaxing safety standards and dose limits? Who will trade the virtue of caution for an accusation of recklessness?

Whether it's industry experts or outsiders in civil society, everyone demands that nuclear plants are absolutely flawless. Never should a glimmer of radiation escape. Endlessly, the most absurd scenarios are fabricated. A regulator sitting at a desk in a nice office conceives an unlikely problem and, eventually, years of research involving state-of-the-art computer simulations show that, yes, the unlikely problem can be prevented or solved, just to be on the safe side. How? By implementing an expensive addition to an already robust safety system.

It's not just that this has very little to do with the reality of a nuclear reactor. These adjustments also fail to reassure. Once you start thinking in absurd scenarios, you will always detect new problems.

Sometimes proponents of nuclear power point out that the public is irrational in its concerns about the health effects of a nuclear accident. But what is 'irrational'? When you've been frightened for decades by disinformation from campaign groups, industry, politicians and the media, isn't fear a logical consequence? Isn't it the obsession with safety in the nuclear industry that has become irrational?

First, we were lied to about how safe a nuclear reactor was. An accident was out of the question, we were told. Then we were lied to about how dangerous a nuclear reactor was. An accident is going to be horrific, we heard. Government officials and journalists readily went along with both narratives. The nuclear industry manufactured the first fable; it was a silent accomplice when the second myth was created.

With an industry like that, who needs an anti-nuclear movement?

Postscript. During his life, Hermann Muller would be showered with awards, but more than half a century after his death in 1967, his reputation took a dent. Investigative work by Edward Calabrese, an American toxicologist, showed that Muller could have known he was wrong.[38]

A colleague of Muller's, one Ernst Caspari, had conducted a similar study to Muller's with fruit flies. Against expectations, he found a threshold dose. Caspari operated with extreme caution, as his academic career was still nascent and his study overturned a widely accepted view. He offered his manuscript to a journal called *Genetics*. The editor, Curt Stern, happened to be a friend of Muller. In the 1920s, they had studied fruit flies together. Stern sent the manuscript to Muller for review.

Responding in a letter, Muller announced that after a quick reading, he noticed that the study contradicted his own data. Because of an upcoming trip, things were rather hectic, Muller wrote, but he would try to read it properly. Of course he would. Not only had doubt been cast on his well-known theory, it had in fact been refuted.

It was November when Muller scanned Caspari's study. Two months later, after returning from his trip, he responded once again. In a letter to Stern, Muller wrote that he had 'little to suggest in regard to the manuscript'.[39] However, at his recommendation one change was made when the manuscript finally appeared in *Genetics*.[40] Only a single sentence had been deleted from the original. That sentence named Muller's error about the threshold dose.

The trip Muller referred to in his letter took him to Stockholm. That was in December. The year was 1946. Hermann Muller had come to collect his Nobel Prize. There, on the stage of honour of science, in possession of a still unpublished study that undermined his findings, Muller would proclaim that with radiation there was 'no escape from the conclusion that there is no threshold dose'.

PART FOUR
OOH!

CHAPTER NINE

Heated

How can nuclear energy combat climate change?

'We are the first generation to feel the impact of climate change, and the last generation that can do something about it.'

— Barack Obama, US president, 2009–2017, remarks at UN Climate Change Conference, Paris, 30 November 2015

It's Monday 20 August 2018, and the summer holidays have ended. But when 15-year-old Greta Thunberg takes her bike out of the garage that morning, she's not going to school. Instead, she cycles to the Swedish parliament building on one of the islands in central Stockholm. She hangs a sign against the granite wall. It reads: 'We kids most often don't do what you tell us to do. We do as you do. And since you grown-ups don't give a damn about my future, I won't either. My name is Greta and I'm in ninth grade. And I am school striking for the climate until election day.'[1]

Ever since Greta could read, climate change had been major news. Things were not looking good for the future of the planet. During the 2015 UN Climate Change Conference (COP21) in Paris, France, global leaders had finally agreed to hold the increase in the global average temperature to well below 2°C, and preferably below 1.5 degrees, compared to the period before the Industrial Revolution. But in the years that followed, signatories to the Paris Agreement weren't doing enough to limit global emissions. They weren't taking climate change seriously. When reading about the steady increase in greenhouse gas emissions, Greta learned that we only had until 2030 to avoid warming on a life-threatening scale. Time was running out. That summer of 2018 had been Sweden's hottest since records began, 262 years before.

When she placed her backpack on the cobblestones in front of the parliament building and sat down, an improbably successful campaign began. Almost instantly, photos of the lone girl's '*skolstrejk för klimatet*', School Strike for Climate, received hundreds of shares on social media. It didn't take long before local journalists took notice, and soon others joined her in her protest – young and old, even school teachers.

Then, on 1 September, an article appeared in the *Guardian*, quoting the teenager: 'I am doing this because nobody else is doing anything. It is my moral responsibility to do what I can.'[2] Greta's message quickly travelled around the world. Weeks later, after the Swedish parliamentary elections, Greta announced she would go on strike every Friday. Hundreds of thousands of students in other countries began following her example.

With her simple appeal to Swedish politicians to reduce greenhouse gas emissions according to agreements already made, Greta Thunberg turned into the world's best-known climate activist. After bursting onto the scene, she was showered with awards from universities, campaign groups and even the Pope. *Time* magazine named her 'Person of the Year', the youngest ever. More than once, she was nominated for the Nobel Peace Prize.

A year later, when Greta was invited to be a guest speaker at the 2019 Climate Action Summit at the UN headquarters in New York City, to which she travelled in a sailing boat, she was undeterred. 'This is all wrong,' she began emotionally. 'I shouldn't be up here. I should be back in school, on the other side of the ocean. Yet you come to *us*, young people, for *hope*?! How dare you!'[3] It was clear she was taking climate change seriously.

She has every reason to. Every new study by the UN's Intergovernmental Panel on Climate Change (IPCC) causes greater alarm. Out of sight to most of us, the vast Antarctic ice sheet is melting faster than ever, causing sea levels to rise dangerously. Due to the retreating ice, it's raining more often on Greenland's ice sheets, and wet snow is less capable of reflecting the sun's rays. Meanwhile, scientists are struggling to find ice floes thick enough for their measuring instruments.

If we don't act now, experts say a 2-degree rise can hardly be avoided and we can expect it as early as the middle of this century. With such warming, rainforests, coral reefs and other ecosystems will be severely damaged, and heat waves and floods may become more frequent.

To stress the urgent need to tackle climate change, some activists prefer to speak of 'climate crisis' or 'climate emergency'. We need to do all we can, and we need to do it immediately.

That's exactly why nuclear power is now back on the table. With nuclear fission, nothing gets burned, so no greenhouse gas emissions like carbon dioxide (CO_2) are emitted. Established organisations of climate scientists and energy experts are therefore clear about the role nuclear plants should have in climate policy. The IPCC recognises nuclear energy as a climate solution and expects its share of the energy mix to increase.[4] The United Nations Economic Commission for Europe (UNECE), which promotes economic cooperation, argues that climate targets will not be met without scaling up nuclear power.[5] The International Energy Agency (IEA) urges governments not to close nuclear plants, but to build new ones.[6]

When Svante Arrhenius began calculating the impact of CO_2 on the climate, he was suffering from depression. It was more than just the general pessimism bubbling up among European intellectuals and artists on the threshold of the twentieth century; his marriage was failing and he didn't know what to do with his life. Arrhenius, a physicist who lectured at the Stockholm University College, sought distraction. So he started thinking about carbon dioxide...

In the nineteenth century, it was known that CO_2 absorbs infrared radiation. Without it, the Earth would be a lot cooler. But, Arrhenius wondered, what happens when the concentration of CO_2 in the atmosphere changes? Could an ice age occur if there's less of it? Would the Earth warm up if there's more of it? For months, he made calculations with pencil and paper. Soon, the aimless thirty-something brightened up: he had discovered something new.

With a doubling of the percentage of CO_2 in the atmosphere, he concluded, the Earth could be 4 degrees warmer. If it was to increase fourfold, the temperature would rise by 8 degrees. Therefore, CO_2 acts as a regulator in the planetary thermostat.

Arrhenius' study appeared in 1896,[7] though hardly anyone showed an interest as most natural scientists were under the spell of the mysterious X-rays and something that would soon be called 'radioactivity'. Undeterred, in 1903, the year Arrhenius received a Nobel Prize in Chemistry for other work, he explained his formulas in a technical book and later in a public version, calling it his 'hot-house theory'.[8]

Still, attention was scarce. With thousands of years to double the concentration of CO_2 in the atmosphere, his theory just didn't seem very urgent. As far as colleagues took note of his work, they considered it irrelevant. Nor did his predicted outcome fit the prevailing worldview. How could a puny creature like man affect something as all-encompassing the climate?

How? Well, for example, by burning a lot of coal, Arrhenius argued. As his contemporaries knew, CO_2 is released whenever coal is burned. And the steam engines that powered machines, locomotives and ships all burned coal. More steam engines = more coal = more CO_2 = more warming. Should a global industry ever get going, Arrhenius reasoned, this warming might occur within a few centuries.

But very few at the time saw the cause for concern. The world population stood at 1.5 billion, and many of these people were still living as peasants had in the Middle Ages. Certainly in his native Sweden, with relatively few hours of sunshine per year, rising temperatures were not perceived as a problem.

Still, this wasn't enough to put a dampener on Svante Arrhenius. He found a new wife, discovered that plants grow faster thanks to CO_2 and, besides, a new century had begun. Progress was sweeping through Europe. The modern world was taking shape. What was there to worry about? His hot-house theory faded into obscurity.

In 2003, a year in which tens of thousands of Europeans succumbed to an extreme heatwave, a girl was born in Stockholm with the same surname as Arrhenius' mother: Thunberg. The girl, Greta Tintin Eleonora, is a distant relative of the Nobel laureate. Her father, the actor Svante Thunberg, is named after him.

In *Our House Is on Fire: Scenes of a Family and a Planet in Crisis*, the Thunbergs forgive Svante Arrhenius for misjudging the speed of global warming. After all, how could he have predicted that generations after him would burn fossil fuels with such fervour that one might wonder whether this stuff 'should have been kept in the ground'?[9]

In the ground, however, fossil fuels are of little use. Dig them up, light a flame underneath them, and miracles start to happen. By burning coal, a steam turbine cranks the generator that produces electricity, but we also use coal to make steel and iron. Oil enables cars, planes and container ships to move, and is used to make plastic and other synthetics. Natural gas heats our homes and our water, and it enables us to make fertilisers. Fossil fuels are used to produce almost any manufactured item you can think of: cement, aluminium, electronics, glass, glue, furniture upholstery, bicycle tyres, asphalt, paper, ink, you name it.

This tells us two things. One, it's not without reason that we started taking these resources out of the ground. Two, it won't be easy now to leave them there.

That's bad news. Every time we burn fossil fuels, greenhouse gases such as carbon dioxide, nitrous oxide, methane and ozone are released. In the atmosphere, these gases absorb the heat radiated by the Earth's surface. When they accumulate, the temperature on our planet gradually rises. Around 50 billion tonnes of greenhouse gases are emitted globally every year.[10] The goal is zero.

It's understandable that Greta Thunberg, and many others like her, worry whether that goal will ever be achieved. It's a monumental task.

That's why energy production must be shaken up first. Above all, we should switch to using electricity. Instead of gas boilers, switch to heat pumps. Instead of gas cookers in the kitchen, induction hobs. Instead of petrol guzzlers at the filling station, electric cars at a charging station. Electricity can also be used to produce hydrogen, which can serve as a substitute for natural gas in industry. It is electricity, therefore, that is seen as the workhorse of the energy transition.

Fortunately, we don't need to burn fossil fuels to produce electricity. However, the alternatives often depend on what nature has to offer. Countries blessed with mountainous landscapes and flowing rivers or high-altitude lakes, such as China and Canada, already generate and use a lot of hydropower. Geothermal energy works in volcanic hotspots with high seismic activity, as happens to be the case in Iceland and the Philippines. Solar and wind, on the other hand, are available everywhere, albeit not in equal quantities and not every moment of the day. Countries deploying thousands of solar panels and wind turbines include Denmark and Germany.

It turns out that nuclear fission is the only energy source that doesn't depend on geographical location or the whims of the weather gods. Any country can deploy nuclear power, as long as it meets a set of conditions, including willingness to submit the plants to international monitoring bodies. So far, France is leading the way. It has by far the largest share of nuclear power in its electricity mix, at up to 70 per cent, followed by Hungary (44 per cent) and Slovakia (41 per cent).

Although no greenhouse gases are emitted when electricity is produced through any of these ways, emissions *are* released when resources are mined and materials produced for manufacturing installations, or during fuel transportation and waste treatment. It's therefore worth examining the average greenhouse gas emissions during the entire lifecycle of each of these low-carbon energy sources.

The IPCC states that coal leads the list with 840g (1.9lb) per kWh. Natural gas sits at 490g (1.1lb), biomass at 230g (0.5lb). Everything else follows at a distance, with rooftop solar panels at 41g (1.4oz), geothermal 38g (1.3oz), hydropower 24g (0.8oz), and both nuclear and wind down the list at 12g and 11g (0.42oz and 0.39oz) respectively.[11]

Nevertheless, Greenpeace maintains that with regard to carbon emissions, there are 'no significant savings' through nuclear power compared with fossil fuels.[12]

Producing electricity without generating greenhouse gases is a challenge. The good news is: we already know how to do it, and a number of countries with modern electricity grids are seeing decent success.[13]

For example, Norway and Paraguay are doing it almost entirely with hydropower. In countries like France, Switzerland and Sweden, they're using a combination of hydropower and nuclear to produce at least three-quarters of their electricity, with the rest coming from fossil fuels and a bit of solar and wind. This is also the formula for success in the Canadian provinces of Quebec and Ontario, which have regional economies larger than those of many European countries.

In 15 years, France turned on 56 nuclear reactors and, along the way, electrified everything from home heating to high-speed railways. Sweden, Greta's home country, reversed its use of fossil fuels even faster.[14] Again, the rapid construction of nuclear reactors in the 1970s and 1980s was enough.

However, there's still a strong preference for solar and wind. These renewable sources possess an irresistible appeal: they bring us closer to greater harmony with nature. In the 1960s and 1970s, solar panels and wind turbines were promoted for energy independence; now they should ensure a fossil fuel-free future.

And so since the 2000s, climate policy has focused not so much on phasing out fossil fuels, but on expanding the use of renewables. Success was measured not by the drop in CO_2 emissions, but by the increase in the share of renewables in the energy mix. Policy was not based on following the example of countries that had succeeded in keeping emissions low, but on complex computer models with future scenarios full of assumptions. Also helpful was the decision to neglect the substantial CO_2 emissions of biomass, allowing for a renewable energy source to produce electricity when wind and solar weren't available.

Incentives were introduced to give new life to old-fashioned energy sources. Entire industries were built. Ever larger wind turbines (with a tip height now exceeding 250m/800ft) were erected, mainly around villages or along the coast. Solar panels took over farmlands. Wood pellets were transported by ship for burning in coal-fired power stations. Governments were generous in providing subsidies, credits, loans and tax benefits.

Promises were made that sounded vaguely familiar. Some said that renewable energy would not just be cheap, nor even 'too cheap to meter', as was once claimed about nuclear power, but 'free', as one CEO of an electricity supplier claimed.[15] In the words of Jeremy Rifkin, an American visionary who is said to have written the blueprint for Germany's climate policy: 'The sun and the wind will never send us a bill.'[16]

Energy companies, however, did send in their bill. In countries where solar and wind made up an increasingly large part of the energy mix, costs soared. In 2021, within Europe, the Danes and the Germans paid the most for their energy, thanks to green levies and fees.[17]

Expensive or not, it remains one big, uncertain experiment. And once you embark on an experiment, you want it to succeed. And so, one fine day in 2021, Austria's minister of economic affairs asked her colleagues in the European Union if her country could perhaps be excused for its poor performance in reducing emissions. She explained:

> We've made a clear statement that we do not want any nuclear energy. That's a principle which is enshrined in our Constitution. We should not therefore have a situation where all countries are treated the same, whether they use nuclear energy to reduce CO_2 emissions or not. We're not in the same position, we shouldn't be treated the same way.[18]

Just as some countries don't like nuclear, others don't like coal. The UK, for example. For a long time, a whole swath of politicians in Westminster were annoyed by mine workers striking for better pay, causing them to decimate rather than resuscitate an industry that, by the 1980s, was languishing under Prime Minister Margaret Thatcher.

Let's compare the UK with Germany. In both countries, about a third of all electricity comes from renewables. During the process of closing down their nuclear plants, the Germans expanded solar, wind and biomass, opened coal plants in 2013 and 2015, and re-activated several already closed ones in the years after. They virtually razed the ancient Hambach forest to open up a new coal mine and dismantled a wind farm to expand the nearby lignite (brown coal) pit.

Since 2013, the British, on the other hand, have closed a dozen or so coal plants; the last is scheduled to shut down in autumn 2024. Well before this, they had installed gas-fired power plants, as well as biomass plants and wind farms, while construction began on a prestigious nuclear reactor at Hinkley Point, the first in what is to be a series. As a result, in the UK, the decline in CO_2 emissions from electricity since 1990 was three times faster than in Germany.[19]

The German government declared the *Atomausstieg* (nuclear phase-out) in 2005. One by one, nuclear reactors closed, with the last ceasing operations in 2023. Wind turbines rose up in rural areas. Solar panels appeared on rooftops. The person moulding this energy transition into policy plans was then-chancellor Gerhard Schröder. He'd grown up in a working-class family and rose to become head of the Social Democratic Party, before being elected as leader of his country. Just as he was heading for election defeat in 2005, he was asked what he wanted to do after his political career. Schröder said: 'Make money.'[20]

And so it was. Less than three weeks after the people voted Schröder out, Vladimir Putin called. He asked if 'Gerd' wanted to work for Nord Stream, the new pipeline bringing Russian gas to Germany, which had been negotiated with the Greens in the latter days of Schröder's coalition government. Well, sure. Schröder foresaw a golden future for Russia's oil and gas industry, or at least for himself. He had himself appointed chairman of Nord Stream's supervisory board and later held board positions at Gazprom and Rosneft, Russia's state-owned gas and oil companies.

The man who, as chancellor, had decided to end nuclear power in Germany, saying the future belonged to renewable energy, thus became a lobbyist for fossil fuels. He reportedly pocketed around €1 million (£850,000) a year from various Russian energy companies.

In February 2022, Putin's army invaded Ukraine. Suddenly, Europeans became aware that quite a lot of their energy is supplied by Russia. In 2021, the European Union imported almost 30 per cent of its oil from Russia, as well as more than 40 per cent of its natural gas and almost 50 per cent of its coal.[21] For these fossil fuel imports, the EU paid some €270 million (£230 million, or $292 million) every day to the Russians, even as they were dropping bombs on schools and hospitals in Ukraine.

But if there's no other source of reliable electricity, you need fossil fuels. And countries that close down nuclear plants need coal or natural gas. That's why it's such a waste to close them.

Sometimes, closure is said to be unavoidable: maybe the nuclear fleet has aged or the reactors have become 'too old'.

However, nuclear plants are constantly being upgraded and monitored to meet the latest safety requirements. In many cases, parts have been replaced. Extending the lifespan of a nuclear plant beyond what was initially agreed can be a political challenge, but it's not really a technical one. In 2021, a group of experts at the US Department of Energy and the Electric Power Research Institute concluded that there are 'no technical limits' to keeping existing US reactors – all well into their fortieth year of operation or more – running for at least another 40 years.[22] The IEA looks at it another way. 'Despite recent declines in wind and solar costs, adding new renewable capacity requires considerably more capital investment than extending the lifetimes of existing nuclear reactors.'[23] In other words: keeping a nuclear plant open is the cheapest and fastest way to produce a lot of low-carbon power.

In 2014, Germany's minister of foreign affairs wrote a letter to Sweden's prime minister, asking if he and Vattenfall, the state-owned energy concern, would please continue to invest in two of Germany's lignite mines. These investments were necessary, the minister wrote, because 'we cannot simultaneously quit nuclear energy and coal-based power generation'.[24]

In Belgium, which had also embarked on a course to phase out nuclear, Energy Minister Tinne Van der Straeten had long championed fossil fuels. As the founder of law firm Blixt, she had been serving virtually 'all the major players in the electricity sector and industry' since 2010, according to research by a Belgian business newspaper.[25] In the press, Van der Straeten, known to the public as a former figurehead of the country's Green political party before setting up her law firm, repeatedly stressed the benefits of gas plants. In an interview she argued that these can be 'quickly switched on when there's not enough sun and wind' and can be 'shut down just as quickly as soon as there is sufficient sun and wind'.[26] Elsewhere, in a commentary piece, she praised gas plants as 'clean, efficient and flexible'.[27]

Van der Straeten's business partner, Tim Vermeir, who'd previously worked for a subsidiary of Gazprom and Eni, one of the biggest players in the natural gas industry,[28] advised the government to invest 'in new or improved natural gas power plants'.[29]

When she returned to parliament in 2019, Van der Straeten arranged hundreds of millions of euros as a 'support mechanism' for the construction of gas plants.[30] Then, in 2021, as minister, she organised even more financial support via the European Commission.[31]

The realisation that a choice for renewables is also a choice for fossil fuels had already dawned on companies in that sector. Over the years, they sent these messages out into the world: Shell posted on X (formerly Twitter): 'No sun? No wind? No problem, natural gas has it covered. See why #natgas is a great partner for renewable power sources.'[32] A slogan from BP read: 'Our natural gas is a smart partner to renewable energy.'[33] An ad from Norway's state-owned Statoil, now Equinor, claimed: 'Natural gas – the perfect partner for renewables.'[34] And the International Gas Union, the industry lobby group, published a report entitled 'Natural Gas: A Partner for Renewable Energy.'[35]

Whether natural gas is a great, smart, perfect or just average partner, solar panels and wind turbines cannot work on their own. Gas plants were necessary to counteract the erratic nature of sun and wind, due to their ability to be ramped up or down. And with big budgets for advertising and lobbying, companies like Shell and Exxon-Mobil linked natural gas to clean energy.

It worked. Politicians, journalists, experts and even climate activists started referring to natural gas as a 'transition fuel'. It would now be tolerated as an intermediary in climate policy, on the way to a completely clean and emissions-free energy system. Under pressure from Germany in particular, the

European Commission decided in 2022 that natural gas would be admitted to a so-called taxonomy of 'sustainable investments' supporting climate targets.

While fossil fuel companies are honest about the need for renewables to work together with natural gas, Green politicians aren't. The Belgian Green Party, for example, firmly states: 'We are moving away from fossil fuels',[36] and we will 'step by step get rid of polluting fossil fuels'.[37] On her own website, Minister Van der Straeten writes that the ambition is to 'phase out' fossil fuels.[38] With that in mind, isn't it odd for Belgium to start building gas plants?

In the media, Van der Straeten constantly assures us that these are only 'a few' gas plants, which are needed 'only temporarily'.[39] We need them 'as an intermediate step', she says,[40] 'as a transition'.[41] On television, she compares the energy transition to a home renovation: 'First, it gets worse before it gets better.'[42]

But once natural gas is chosen as an intermediate step, it will be difficult to get rid of, experts warn. The scientific journal of research in renewable energy unequivocally labels that choice a 'blunder'.[43]

The decision to close nuclear plants and build gas ones instead means that Belgium will generate significantly more electricity from fossil fuels in 2030 than it does today. Whereas that share was 38 per cent in 2018 – lower than the European average – it will reach almost 60 per cent by the end of this decade if nuclear power is finally phased out, according to a 2020 study. By then, the country will languish somewhere at the bottom in Europe, among the biggest polluters.[44]

And so, what just two nuclear plants had prevented for decades is happening in Belgium after all, with the help of the Greens: a modern country's electricity supply is falling into the hands of a dirty old technology that pollutes the air and heats up the Earth.

What are we supposed to make of this? Environmental organisations are certainly not manning the barricades for coal and gas, but their attack on nuclear power does curtail a formidable competitor to fossil fuels that can provide the same services *without* greenhouse gases or air pollution.

If a nuclear plant closes somewhere, the lost capacity is nearly always entirely replaced by fossil fuel-based facilities. That's not because the environmental activists want it to be so, but because their own preferred solutions – wind turbines and solar panels – simply do not provide reliable power and therefore cannot function as replacements. Indeed, their solutions only work when fossil fuel power plants can generate at least as much.

Thus, opponents of nuclear power inadvertently do the work of the fossil fuel lobby. Multinationals like Shell, BP and Equinor are happy to see anti-nuclear activists eliminating their competitors. Opponents of nuclear power have turned out to be the useful idiots of the world's most polluting industry.

In the early years of the global environmental movement, climatology was a small, young field without much prestige. Between themselves, climate scientists speculated about the ever-increasing concentration of greenhouse gases in the atmosphere. Most were cautious. Some wondered whether the use of fossil fuels should be reduced. They received little response.

A first official warning was issued in 1971. Experts from 14 countries claimed that humankind could warm the Earth. If that were to happen, the consequences would be severe. Their *Study of Man's Impact on Climate* was compulsory reading for participants at the first United Nations environmental conference in Stockholm the following year.[45] It didn't gain much interest.

A few years later, the National Academy of Sciences predicted that if things continued as they were, global temperatures could rise to catastrophic levels.[46] It was now 1977. This time the warning didn't go unnoticed. 'Scientists Fear Heavy Use of

Coal May Bring Adverse Shift in Climate', read one front-page headline in the *New York Times*.[47] *Business Week* wrote that CO_2 'might be the world's biggest environmental problem, threatening to raise the world's temperature'.[48]

The question soon arose as to what this might mean for energy production. Such academic reflections became relevant when, a little later, a new oil crisis arrived. The Iranian revolution brought unrest to the region and the price of oil doubled. Now what?

Immediately, US president Jimmy Carter responded to the dire situation. He put on a jumper and turned the thermostat down to 18 degrees Celsius (for a long time, his wife would complain about finding herself so cold, despite being the First Lady). In a much-debated television speech on 15 July 1979, Carter lashed out at the American way of life with its worship of 'self-indulgence and consumption'.[49] He argued for a 'change of course' and advocated energy conservation. He also said he would replace foreign oil with homegrown coal, 'our most abundant energy source'.

Wait, coal? Was that really such a good idea given all those carbon emissions? The nuclear industry spluttered. If coal plants were the problem, shouldn't nuclear plants be built instead?

Well, no, said Carter. During his presidential campaign, he had promised an end to the development of nuclear power. In a high-profile meeting, he had described nuclear plants as 'bomb factories'. His promise had also made the front page of the *New York Times*.[50] Incidentally, the event had been paid for by the Rockefeller Foundation and, according to the article, the address Carter had given was discussed with 'a group'. Seven names followed and all had connections with the Rockefeller family. And now, as elected US president, Carter had decided that nuclear power was no longer required.

Back to Carter's televised address. Referring to efforts made during World War II, the president boasted that he would 'mobilise American determination and ability to win the energy war'. The future, he said, belonged to solar power. Carter had 32 solar panels installed on the roof of the White House. Furthermore, he promised to prepare legislation to ensure that, by the year 2000, the sun would provide no less than 20 per cent of all energy for the US.

In 2000, the share of solar power in the supply of US electricity was a mere 0.01 per cent.[51]

In the intervening years, evidence was mounting that climate change was at least partly due to the use of fossil fuels. The computer models might still not have been perfect, and most experts expressed caution, but there was no escape.

Then, James Hansen appeared on the scene. Back in 1981, the NASA climate scientist had already made the front page of the *New York Times* with his prediction that the evidence of global warming would be visible within 20 years.[52] But now, on an extremely hot June day in 1988, at the beginning of a summer full of parched fields, forest fires and a violent hurricane, Hansen spoke to the US Congress. He sat somewhat uncomfortably at the microphone, but his message was crystal clear: 'The greenhouse effect has been detected, and it is changing our climate now.'[53]

Afterwards, it became apparent that Hansen was not quite used to leaving the world of computer models and entering the media spotlight. When confronted with a question about scientific doubt, he said, rather edgily: 'It is time to stop waffling so much.'[54] For a long time after, he would leave the interviews to his colleagues.

Once again, we pull out the front page of the *New York Times*: 'Global Warming Has Begun, Expert Tells Senate'.[55] Journalists were also beginning to take note in Europe. According to one newspaper article, Hansen had 'sounded the alarm with such authority and force that the issue of an overheating world has suddenly moved to the forefront of public concern'. As the headline goes, there's 'no turning back' from the greenhouse effect.[56]

Politicians could no longer ignore the issue. The next year, UK Prime Minister Margaret Thatcher, speaking at the United Nations General Assembly, said:

While the conventional, political dangers – the threat of global annihilation, the fact of regional war – appear to be receding, we have all recently become aware of another insidious danger. It is as menacing in its way as those more accustomed perils with which international diplomacy has concerned itself for centuries. It is the prospect of irretrievable damage to the atmosphere, to the oceans, to earth itself.[57]

Thatcher, who was trained as a chemist, was known as a free-market advocate of deregulation and privatisation. But there was no sign of this on the stage that day. Instead, the 'Iron Lady' made an impassioned plea for international cooperation and binding agreements on reducing greenhouse gas emissions.

Thatcher's speech fell on a historic day, and not just because Svante Arrhenius' theory had been recognised on the supreme political stage as a problem clamouring for a solution. That day in November 1989 was the last day of the Berlin Wall. Now, with the communist danger gone and the Cold War at an end, a new enemy had emerged – one that threatened all people in all countries. It had the potential to become a battle that would unite the whole world in a common goal: ensuring the danger of disruptive climate change was averted.

Over time, the fight against a warming planet gained more and more allies. Government leaders made commitments to lower CO_2 emissions. Climate scientists put in such effort that they were awarded the Nobel Peace Prize. Environmental activists saw in climate change a new reason to warn of impending doom. Inventors worked on smart solutions. Executives of corporations made pledges to behave better. Pension funds withdrew from the fossil fuel industry. Pop stars and Hollywood actors voiced their concerns about the climate.

Those who had dared to deny global warming, doubt its severity or question climate policy were forced to eat their words.

Yet there was something weird going on. While climate change was gradually becoming the focal issue in the environmental movement, certain groups within that movement continued to oppose a leading source of low-carbon energy. Did they really consider their own dislike of nuclear energy to be more important than the collective fight against climate change?

As if global warming didn't exist, action groups continued to do their best to close nuclear plants and prevent the construction of new ones. As recently as 2019, a senior expert in energy policy at Greenpeace wrote: 'Sabotaging nuclear is a vital part of any successful attempt to save the climate.'[58]

It puts us today, more than half a century after the first serious warnings about global warming, in the strange situation of having heard for a long time now that nuclear power is 'too late' to help us do anything about climate change. But: who is too late here? Did the activists themselves perhaps waste time? Could they not have been quicker to recognise climate change as a serious problem and accept nuclear power as part of the solution?

Once such questions pop up, new ones keep coming. Do they really care as much about the climate as they say they do? How bad should we feel about the climate anxieties of the environmentalists, when they allowed the anti-nuclear activists in their midst to obstruct a solution they didn't like? Isn't climate change at least to some extent the result of their opposition to nuclear power? If so, isn't it time, then, that we started holding them accountable?

Things were bound to go wrong at some point. At a climate rally held in September 2021 in Berlin, Britta Augustin was standing close to the stage where Greta Thunberg had just been speaking. Tens of thousands of people were gathered on the square in front of the Reichstag building. Augustin, a maths teacher and mother of three, held up a cardboard sign. 'Kernkraft gegen Klimawandel!', it read. Nuclear power against climate change. Videos circulating on social media showed what happened next. A man came from behind, grabbed her by the wrist and snatched the sign away. When Augustin held on to it, the man dragged her to the ground and tore up the board. Bystanders clapped and cheered.[59]

A month and a half later, at a climate march in Amsterdam, a group of nuclear advocates, dressed in inflatable polar bear suits, danced happily to the music and took photos with children. Not everyone appreciated it. Later in the day, an angry woman punctured two of the suits.[60]

These are just two of many incidents in which pro-nuclear climate activists have been attacked by anti-nuclear activists.

This split within the environmental movement has been years in the making. The debate plays out mainly on social media. Here, outside the narrow opinion spectrum of mainstream media, outspoken supporters of nuclear power love to share their views. They won't hear a negative word about their climate solution of choice. They find their critics ill-informed, arrogant and blinded by ideology. They maintain that these critics, who often promote renewables, are only considering the advantages of solar and wind but *not* the disadvantages. Of course, the nuclear advocates consider themselves quite rational and informed about everything.

The same accusations can be heard in reverse. Those who are against nuclear power consider the loudest advocates to be smug and obsessed with their favourite technology, of which they see only the benefits. *Sigh, if only they were more rational and informed about everything...*

Proponents of nuclear power, however, have noticed a disturbing weakness in their adversaries. The opponents of nuclear energy seem to dream of all kinds of solutions that *do not yet exist.* They hope for large-scale storage of excess green power in super-batteries or hydrogen, or the introduction of a global carbon tax, or a worldwide electricity grid that intelligently matches supply and demand. Maybe, who knows, one day these things might come to fruition, but they don't seem to have much appreciation for, or even interest in, an existing, proven technology that in the real world provides societies, including some of the world's largest economies, with low-carbon energy.

Both sides love calculations – that is, as long as they confirm their own beliefs. Exactly what the maths shows is not necessarily of much use. If you assume drastically lower energy consumption, batteries that aren't yet available on a significant scale and a gigantic fleet of hydropower plants for which there's no suitable place on Earth, you'd find that – *yes!* – 100 per cent renewable energy is possible.[61] Similarly, you can calculate, as James Hansen did, that 115 nuclear reactors will have to be built every year until 2050, at which point the entire global electricity supply will be low-carbon.[62] But, of course, just as it is inconceivable that the world will run on renewable energy, it will not run entirely on nuclear power either.

Yet there is an important difference in perception. Those who make such calculations for renewables are considered idealists, while those who make them for nuclear seem suspect. And so, one day, James Hansen came to be branded as a kind of climate change denier.

The accusation didn't come from just anyone. Naomi Oreskes is Professor of the History of Science at Harvard and has written about climate change deniers in her acclaimed book *Merchants of Doubt*. So what is this 'new, strange form of denial' of which she says Hansen is guilty?[63] He has implied that renewable energy sources cannot meet all the world's energy needs.

For the record, in the article to which Oreskes refers, Hansen wrote that climate policy 'must be based on facts and not on prejudice'.[64] He also argued for deploying all clean energy sources and increasing investment in both renewables and nuclear, noting that it makes no difference to the climate whether the avoided greenhouse gases are attributed to a wind turbine, a solar panel or a nuclear plant.

Apparently this meant that Hansen was a 'denier'.

It seems that those who like nuclear power irritate those who don't. Asked by her mother what the most annoying thing is about her critics, Greta Thunberg mentions those who want to talk with her about nuclear power. She complains that they don't seem to know anything and simply ask her what she thinks of nuclear power, and then 'smile as if they have solved all the world's future problems on their own.'[65]

A conversation unfolds in the Thunbergs' home, which is recorded word for word in *Our House Is on Fire*. Greta says lots of fossil-free energy is needed, sooner rather than later, so investments are needed to get the fastest results at the lowest prices. Since nuclear plants are expensive and take such a long time to build, the best alternatives are wind turbines and solar panels. After all, these are much cheaper and can be deployed much faster. Greta concludes that she can't fathom why some people keep talking about nuclear power.

Why do they? Because, as we have seen, nuclear plants compete with other power plants. Nuclear plants make coal plants redundant. They can also be used to produce hydrogen, making natural gas plants redundant.

Precisely because power stations using fossil fuels can be replaced with nuclear plants, existing energy infrastructure can be maintained. That's an important reason why the power inside atoms not only makes the transition to a low-carbon society possible faster than with renewable energy, but also cheaper.

Yet many people say nuclear power is expensive. For example, Frans Timmermans, who until 2023 was the Eurocommissioner tasked with making the European Union climate-neutral by 2050, says that a country is free to decide whether it thinks it needs nuclear power, but he wants to make clear that nuclear is 'very, very expensive'.[66]

A high price tag sounds only logical. After all, nuclear power has both the advantages of fossil fuels (constant, reliable) without the disadvantages (pollution, scarcity), and the advantages of renewables (clean, low-carbon) without the disadvantages (weather dependence, land use). Also, a nuclear plant needs to be maintained with good care to ensure the highest levels of safety. Something like that, one might think, must cost a fortune.

The reality is much different. In most countries, power from nuclear plants is cheaper than power from coal or gas. This is certainly the case in countries where energy is a public utility that has not been privatised and is still operated largely by a governmental body, although sometimes not in countries with large fossil reserves. According to guidelines from the International Atomic Energy Agency, the price of nuclear power includes insurance against accidents, plus reserves for nuclear plant decommissioning and nuclear waste disposal.

Power from solar panels and wind turbines can be cheaper – at least when the sun shines and the wind blows. If the elements don't deliver, the costs are borne by the power plants that come to the rescue. All that adjusting and switching due to erratic production from weather-dependent sources makes the electricity supply more expensive. The higher the share of solar and wind energy, the higher the costs. We see this reflected in energy bills and tax charges.

A new nuclear plant is only shut down for a few weeks a year for scheduled maintenance and fuel change. Such a plant will easily make it to the end of this century. But building one is expensive and considered a risky investment because it only pays off in the long term – and only if politicians don't pull the plug. A new wind turbine, on the other hand, produces power at best half the time, has to be painstakingly fitted into the grid, and is likely to be dismantled before 2050. But a wind turbine doesn't cost much and delivers quick profits. It's the kind of short-term thinking that's rewarded in the financial world.

As energy pundit Mark Nelson likes to say: whereas solar panels and wind turbines are cheap ways to make energy expensive, a nuclear plant is an expensive way to make energy cheap.[67]

But most critics of nuclear power appear not to be really interested in economic analyses. Therefore, they employ selective arguments. For instance, they may point out that 'the market' doesn't show much interest in financing nuclear plants, even while they may usually be critical of the free market. Also, the lack of interest from the financial world in solar and wind energy, prior to government incentives, didn't make them argue against renewables.

Meanwhile, the critics seem to believe that the cost of building nuclear plants is borne by taxpayers. Usually that's not the case. Because governments have been shown to be unreliable partners who may decide to cancel construction plans along the way or close down well-functioning nuclear plants prematurely, banks only loan money to investors against high interest rates. Paying off this interest is by far the largest cost item. It could be brought down substantially if a government would only provide a guarantee.

In all this bickering over whether low-carbon power should come from nuclear plants or renewables, there is a laughing third party: the fossil fuel industry. Fossil fuels make up the common enemy, but in the fight between nuclear proponents and nuclear opponents, they remain mostly out of the picture.

Perhaps the problem for Greta Thunberg and her fellow activists is that nuclear plants make something else redundant besides coal and gas plants: wind and solar farms. Why sprinkle wind turbines and solar panels across the landscape, requiring all sorts of additional cables so their intermittent power production can be fitted into the grid, along with a need for all sorts of backup systems, when nuclear plants can be build on industrial sites and run on their own, while demanding a far smaller footprint?

Some are honest about the cannibalising effect of nuclear plants. In the Netherlands, John Hontelez organised protests against nuclear power in the 1970s and 1980s. Decades later, he's affiliated to a Dutch wind energy cooperative, and still not convinced that nuclear is a good idea. 'After all, you won't turn a nuclear plant on quickly when there's no wind, or off when it's windy,' Hontelez says. 'So the nuclear plant will keep running. That way wind and solar won't have a chance.'[68]

He's not quite right. Nuclear plants are perfectly capable of following demand; they have been doing it that way in France for a long time. However, it's a lot cheaper for society if a nuclear plant operator doesn't have to bother about the weather and can let the thing run continuously.

This begs the question of why power from solar and wind should be prioritised. It led British environmentalist George Monbiot to a few observations:

> We should ask ourselves what our aim is. Is it to stop climate breakdown, or is it to engineer the maximum roll-out of renewable power? Sometimes it seems to me that greens are putting renewables first, climate change second. We have no obligation to support the renewables industry – or any other industry – against its competitors. Our obligation is to persuade policy-makers to bring down emissions and reduce other environmental impacts as quickly and effectively as possible.[69]

A call for nuclear power out of climate concerns seems very contemporary. It isn't. Suggestions that clean nuclear plants could play a useful role in replacing dirty fossil fuel plants go way back.

In her speech to the UN General Assembly in November 1989, Margaret Thatcher announced that her government would turn to 'non-fossil fuel sources, including nuclear'. For nuclear power, Thatcher added, 'despite the attitude of so-called greens, is the most environmentally safe form of energy'.[70]

Thatcher was not the first to mention that nuclear power should be considered to avoid global warming. Two days before her historic address, at a symposium on energy in Chicago, a speech was given in which it was stressed that energy conservation and solar panels could accomplish much, but that we shouldn't assume it would be enough. This was the final sentence: 'To deny [nuclear power] a role in the formidable task of controlling CO_2 is surely short-sighted.'[71]

The speaker was Alvin Weinberg, former director of the Oak Ridge National Laboratory, where he worked on nuclear energy innovation. Back in the 1970s, Weinberg had studied the effects of CO_2 on the climate. Speaking as an expert in the US Congress in 1977, he stressed that if the 'carbon dioxide problem' was to become as big as some scientists claimed, the United States needed to be 'in a position to shift quicker than we would otherwise have shifted to the nonfossil sources of energy, and these include nuclear, and they include the renewable sources'.[72]

Earlier, in 1974, Weinberg wrote an article in *Science* in which he urged better knowledge about the climate so that energy policy could be based on sound understanding.[73] Two years later, he argued that the increased demand for energy would lead to higher CO_2 emissions that 'warrant serious concern'. And so, any 'decision made now on the nuclear/non-nuclear issue will have an impact reaching many years into the future'.[74]

But who was listening? Who attended symposia on energy systems when you had wind turbines and solar panels? Who in the environmental movement read scientific journals and papers when science was merely an opinion? Who trusted the talk of Alvin Weinberg, a nuclear physicist who surely must have been a lobbyist for the nuclear industry?

Besides, wasn't it rather suspicious that in the late 1950s it was the nuclear industry that funded the research of Charles David Keeling, who built a measurement point for carbon dioxide in the atmosphere on a volcano in Hawaii? His findings resulted in the famous Keeling curve, which shows higher temperatures at higher concentrations of CO_2.

The Limits to Growth, the legendary 1972 Club of Rome report, featured a graph on the rise of carbon dioxide in the atmosphere. The authors wrote: 'If man's energy needs are someday supplied by nuclear power instead of fossil fuels, this increase in atmospheric CO_2 will eventually cease, one hopes before it has had any measurable ecological or climatological effect.'[75]

Yes indeed. The Club of Rome warned about climate change and recognised the usefulness of nuclear power in the fight against it. But the movement against nuclear did not listen. Nuclear energy was not supposed to be a solution.

Not then.

Not now.

Not ever.

Just before his death in 2006, it dawned on Alvin Weinberg that the anti-nuclear activists at the heart of the environmental movement might be a greater danger than those who deny climate change. Weinberg wrote: 'How can one be *against* nuclear energy if one is concerned about carbon dioxide? To my utter dismay, indeed disgust, this is exactly the position of some of the environmentalists.'[76]

During the 2021 Climate March in Amsterdam, one of the participants, dressed in the jacket of the GroenLinks (Green-Left) Party, carried a simple cardboard sign on which he had written: 'I would rather have nuclear waste than drown'.[77] Because of the urgent threats posed by rising sea levels, he was willing to accept nuclear energy – not as an ideal solution, but as one that was necessary, even though it came with a serious concern.

His dilemma will be recognised by many. Because, yes, all right then, nuclear power can indeed help prevent further warming of the planet. But what about the waste? Nuclear waste is still a big, unsolved problem, right?

Right?

CHAPTER TEN

Hidden Treasure

What should we do with nuclear waste?

'All the wicked of the earth you discard like dross; therefore I love your statutes.'

— Psalm 119:119 (NIV)

On Olkiluoto, an island dotted with pine trees in southern Finland, work is underway on a unique feat of engineering. Here, a complex system of tunnels will be built as deep as 500m (1,600ft) below the Earth's surface. Onkalo is the name of the construction project. All of Finland's nuclear waste is to be safely stored here for hundreds of thousands of years. For, say, eternity.

The granite bedrock consists of solidified magma that formed some two billion years ago. Since then, it has survived a few ice ages during which it remained undisturbed. Rock solid: that inspires confidence when looking for a place to store the radioactive waste from Finland's two nuclear plants, which together provide more than a third of all the country's electricity.

The Finns feel the construction of Onkalo to be a moral duty. Vihreät, the Green political party, supports the project. Articulating the sentiment among party members, environmentalist Tea Törmänen told *HuffPost*: 'They thought, "We have these nuclear plants; we must deal with the waste responsibly".'[1]

The first exploratory drilling began in 2004. If the final touches go according to plan, Onkalo (Finnish for 'cave' or 'hollow') will open in 2025. From then on, highly radioactive nuclear fuel rods, packed in thick copper canisters, will regularly travel underground through an ingenious computerised system. Along boreholes with sharp bends, they will descend further and further into the ground.

Each time a shaft is full, it will be covered by bentonite. This type of clay will fill all the spaces. Bentonite expands when it gets wet, thus preventing water from seeping in. And then, eventually, in a hundred years or so, a layer of concrete will be placed on top and there will be nothing left to add. Finally, the entrance to Onkalo will be sealed up, inaccessible to anyone, ever.

And that's it then. With the opening of Onkalo, Finland is finally relieved of a legacy we have all been saddled with since we started splitting the atom. This will end the complaint that there's no solution to nuclear waste.

But will it? The first doubts are already being heard. Aren't those copper canisters prone to corrosion? Could the groundwater not get contaminated and perhaps flow upwards? How do we point out to our peers in the distant future that there's some extremely dangerous waste buried here? Will things really be all right?

Somehow, nuclear waste gives us the jitters. This anxiety was rather amusingly depicted in *The Simpsons*, the animated sitcom starring a character named Homer Simpson, who works as a safety inspector at the nuclear plant in the fictional town of Springfield. We see the waste as poisonous green liquid sludge in rusting drums, bubbling up against the sides as if alive. The owner of the nuclear plant, the miserly Mr Burns, hides the stuff at workers' homes, in public parks and playgrounds.

To be sure, this is not reality.

For the vast majority of elements found in nuclear waste, radioactivity is extremely low. These waste elements come not only from nuclear plants, but also from hospitals, laboratories and mining. Think of protective gloves and hairnets for staff or filters and scrap from machinery. Before this low- and intermediate-level nuclear waste is incinerated it is collected in stainless steel canisters that are compressed and encased in concrete.

Once such waste was dumped in the ocean. This invited spectacular protests. In 1982, Greenpeace wanted to prevent tonnes of nuclear waste from being thrown into the sea. Activists in tiny inflatables sailed close to the big dumping ship. One of the heavy barrels landed on top of a boat, launching an activist into the sea. From the Greenpeace ship *Sirius*, a colleague shouted into the megaphone: 'Look what is happening! Stop immediately! Are you crazy?!'[2] Thanks in part to footage of this action, a moratorium on dumping radioactive waste in the sea was imposed in 1983. In 1994, this became an international ban.

Few dared to say it, but these barrels of low- and intermediate-level radioactive waste could not do much harm. At the bottom of the ocean, the drums will slowly rust and the contents eventually mix with all the natural uranium and other radioactive minerals under water. Because of the dilution, no plankton will ever notice.

Campaign groups and the nuclear industry alike are making a big deal about it, but this detritus can hardly bear the name 'nuclear waste'. Let's save that term for the spent fuel that comes out of the reactor. These pellets of enriched uranium, packed in rods, must be carefully shielded. For anyone standing next to an unshielded fuel rod fresh from the reactor, the radiation is lethal.

Some 95 per cent of the spent fuel from a nuclear reactor gives off its radiation only very slowly. This proportion includes uranium and plutonium. Only after four and a half billion years has the radioactivity of the most common isotope of uranium halved. It concerns alpha rays, which cannot penetrate the skin. In theory, the slow decay should reassure that the radioactivity is quite weak. In practice, it fosters claims that nuclear waste is dangerous for thousands or even millions of years.

The remaining 5 per cent consists partly of isotopes that give off their radiation within a matter of days or even seconds. This is potentially harmful material, such as iodine-131. The good news is that highly radioactive particles decay fairly quickly. Another small portion are isotopes like strontium-90 and caesium-137 that lose their properties as relevant radiation sources after about 300 years.

What happens to the spent fuel? First, it must be cooled in water. Within a few years, the radioactivity is already greatly reduced. And then? Some countries, like Finland, leave it in the water. After 100 years or so, its radioactivity has decreased by 99.9 per cent.

Other countries take the waste to a reprocessing plant, such as in La Hague, France, where plutonium and uranium are extracted to reduce the volume of the waste and to recycle fissile materials into a new fuel blend. After this process, only 10 per cent of the waste remains, to be poured into glass or concrete. Then, it is stored in impermeable casks, sometimes simply standing outside in the car park of a nuclear plant. These casks can withstand the strangest accidents, the hottest fires and the most extreme natural disasters.

And then, as the radiation continues to decrease naturally, it's ready for final disposal, for instance in the granite bedrock below Olkiluoto.

Deep geological disposal was first proposed by the American Physical Society in 1978. In an authoritative report, this organisation put an end to wild speculation about where nuclear waste should eventually go.[3] Dump it onto a remote island in the ocean? No. Submerge it in a glacier? No. Fire it into space? No.

Burying it deep underground, in granite, salt or clay, emerged as the best option. Technically it was a solved problem. Now, politicians had to make a decision. However, nothing much was happening.

Anti-nuclear groups showed more interest in raising concerns than in exploring solutions. At a multi-day strategy session in 1991 at the Hyatt Regency Hotel in Washington, D.C., Greenpeace and other campaigners decided that nuclear power could be stopped 'by halting development' of waste disposal.[4] Transports to bring nuclear waste to a safe place should be disrupted. Lobbying efforts for ever tighter safety requirements for waste disposal were encouraged.

Anything that could bring about safe disposal of nuclear waste had to be obstructed. Why? Imagine the waste problem was solved. It would mean the end of a powerful argument against nuclear energy.

And so US-based campaigners, preferring to see problems rather than solutions, tried to halt development of the so-called integral fast reactor at the Argonne National Laboratory. This reactor would not only produce significantly less nuclear waste but could even use it as fuel. Their campaign was successful. In 1994, President Bill Clinton pulled the plug.

When, in 2020, the Belgian government announced it wanted to involve citizens in a broad public debate on final disposal of nuclear waste, the Green Party, Groen, objected. It reasoned that final disposal would be 'irresponsible', as it would 'simply shift the problems to future generations'. Therefore, the government should organise a broad public debate. Realising this is exactly what the government intended to do, Groen then argued that such a debate involves 'thousands of generations', and therefore the decision on nuclear waste disposal should be made in the year 2100.[5]

Can anyone follow this reasoning?

And now, after so much effort to obstruct solutions for dealing with nuclear waste, Greenpeace calls it an 'unsolved' or even an 'unsolvable' problem.[6]

If you were a hippie and wanted to escape from the city and live in harmony with nature, Stewart Brand was the man to turn to. Between 1968 and 1972, Brand created *Whole Earth Catalog*, a magazine with advice on how to grow food without pesticides, how to compost, how to build a windmill or a yurt. His publication became a legendary survival guide to a sustainable, off-grid lifestyle, and he himself became an icon. Even when his interests shifted, realising early on the potential of computers, Brand would remain an inspiration to the Green movement.

In 1996, Brand created a think tank, the Long Now Foundation, through which he encouraged people to think about the long term – the *very* long term. Its prestige project is a cuckoo clock that ticks once a year and whose cuckoo appears once every thousand years. To remind us that the human species will be here for several more millennia, Brand likes to write years in five digits. And one fine day in 02002, something happened that would change his views on nuclear power.

That day, Stewart Brand visited Yucca Mountain in the Nevada desert where a repository for the spent fuel from US nuclear reactors was being built. Billions of dollars had already been poured into it. Still, it was far from finished, and the project has since been cancelled.

Before Brand's visit, the Long Now Foundation was asked how to make it clear to our descendants that they should stay away from this site. The briefing initially mentioned that the waste had to be kept safe for 10,000 years. *Ten thousand years*, Brand pondered. *Ten thousand years...*

During the process, safety requirements were increasing. Now, the waste had to be safe for 100,000 years. *No, wait!* some protested, *it should be a million years!*

Then, standing at the foot of an ancient desert mountain by a pretty expensive hole in the ground, Brand realised: this timeframe is what drives people crazy. He perceived that it's an illusion to think we can know what is right over such a long period of time. We can challenge ourselves to *think* about it, but we cannot *act* on it.

A hundred thousand years ago, only a few Neanderthals lived in Europe, a continent still covered by ice. A hundred thousand years from now, humans may have changed beyond recognition, perhaps fused with technology and artificial intelligence. In that case, Brand reasoned, our descendants will probably have 'far more interesting things to worry about than some easily detected and treated stray radioactivity somewhere in the landscape'.[7] And what if our species unexpectedly fell back into a state of poverty, disease, war and ignorance? Well, Brand thought, that source of radiation deep in the ground won't be the biggest concern either.

According to Brand, we should temporarily store the waste for, say, 100 years, so that future generations can still access it. If they want, they can take it somewhere else, or leave it like that, or do something meaningful with it. The man who wanted us to think about the long term realised that the idea of storing something for thousands of years is 'a classic example of the folly of long-term planning'. In his 2010 book *Whole Earth Discipline*, he wrote: 'The more I thought about the standard environmentalist stance on nuclear waste, which I had espoused for years, the nuttier it seemed to me.'[8]

Brand immersed himself in the subject and discovered something else he didn't know: nuclear waste is only a small problem – literally. That's because in a nuclear reactor, only a tiny bit of uranium is needed to produce an amazing amount of energy.

How tiny? A typical nuclear reactor generating 1,000 megawatts, which could supply electricity to a million households for a whole year, might produce 3 cubic metres (106 cubic feet) of high-level waste annually. It would take more than 200 years to fill an average-sized family home.

Suppose all the energy in your entire lifetime had to come from the splitting of atoms. The uranium needed would fit inside a ping-pong ball.[9] With so little going in, little will be going out. The waste produced altogether (which will be a bit larger than a ping-pong ball because of transmutation: the elements made in nuclear fission have a lower density and therefore need more space) would fit into a fizzy drink can.[10] Nearly all of what's in that single can could go into the incinerator. The highly radioactive, non-incinerated portion would not exceed 5g (0.2oz), or the weight of a sheet of paper.

Nuclear waste is compact. As a result, it doesn't take up much space. If we put together all the spent fuel that has been collected from nuclear reactors around the world since the 1950s, which according to the prevailing view should be stored in the Earth's crust, it would fit into a single football stadium. Within the chalk lines of the pitch it would form a block 4m (13ft) high.[11]

In just two hours, the world produces the same amount of electronic waste, and in three minutes, as much household waste as all the world's nuclear plants produce high-level radioactive waste in one year.[12] Whereas that bit of nuclear waste is carefully isolated in water or in casks, such care is not applied to all the other waste that's potentially dangerous to public health and the environment. That detritus is much less strictly regulated and could end up in landfill. Often, such waste goes into the air through chimneys or through pipes into oceans and rivers. Strangely, these mountains of waste, from everyday municipal detritus to toxic liquids generated by industry, hardly meet any resistance.

And while radioactivity in nuclear waste decays according to a natural process and eventually disappears, waste from chemicals, heavy metals and certain plastics never breaks down. Radiation decays faster than many think. After 100 years, the radiation of a barrel of nuclear waste has come down so much that you could safely put it in your living room for an evening and watch a movie,[13] such as the highly recommended internet documentary *How I Learned to Stop Worrying and Love Nuclear Waste*.

Let's take a look at plutonium, which is seen by some as the most toxic substance on Earth. There are plenty of heavy metals and organic substances far more deadly than plutonium that often end up among ordinary waste, such as lead in batteries, mercury in energy-saving light bulbs and cadmium in chocolate. True, these are not large quantities that would cause immediate harm – and that's exactly the point. That little bit of plutonium in nuclear waste is also unlikely to be enough to make you ill.

After exaggerating the risks of even a tiny amount of plutonium, Ralph Nader, a well-known US environmentalist and consumer advocate, was once publicly challenged. Bernard Cohen, a physicist decorated with the highest honours of the American Physics Society, knew that it can't do any harm to carry around a lump of plutonium, and that its radioactivity couldn't penetrate a piece of paper, let alone skin. Cohen also knew that plenty of people have been exposed to plutonium and thoroughly examined. These included workers in nuclear weapons laboratories who had inhaled plutonium dust, but lived in good health well beyond retirement.

Damage also remained limited in human subjects at US hospitals and prisons in whom plutonium was injected, during and just after World War II. The first was Ebb Cade, a construction worker who, as a 'well developed', 'well nourished' and 'colored male', met the requirements for the study. On 10 April 1945, Cade suffered broken bones in a road accident. Without his consent or knowledge, he was administered plutonium. House painter Albert Stevens was another who involuntarily received plutonium into his body. It was so much that he is known to have survived the highest lifetime radiation dose ever: 64,000 mSv.[14]

Of the dozens of people like Cade and Stevens who were given doses of plutonium, not one died from it. They lived on for years – Stevens for more than 20 years, until he was 79 – and their cause of death had nothing to do with plutonium.[15]

Bernard Cohen knew all this. He challenged Nader.[16] He himself would eat as much plutonium in front of the television camera as Nader was willing to consume pure caffeine – you know, that uplifting yet toxic substance in coffee. Wisely, Nader didn't bite.

Or take the waste from other energy sources. At coal and gas plants, substances like mercury and arsenic go straight into the air, causing respiratory damage. When solar panels are dismantled in the landfills of poor countries, they may be burned for the silver and copper wire; then, the toxic fumes from molten plastic and heavy metals like cadmium and lead can contribute to cancer. If only this waste was handled as carefully as that from a nuclear plant, we would be spared considerable environmental damage. The air, soil and water would be much less polluted.

If we were just as afraid of all the other waste that's damaging to our health and our planet, we would have devised as ingenious a detection system for it as we have for nuclear waste. Even the smallest radiation leakage can be detected instantly by highly sensitive equipment that can measure radioactivity down to levels too low to have any discernible impact on humans or the environment. If only we had something like that for particulate matter, sulphur oxides, dioxin, mercury, cadmium, barium, thallium, all those heavy metals that are proven pathogens and even killers.

Putting the risks of nuclear waste into perspective, it turns out that they are insignificant. In fact, altogether, nuclear waste is rather uninteresting. Maybe that's why it's made to seem so complicated. So complicated, that any confrontation with the facts feels unreal.

It doesn't help that the nuclear industry has no qualms in maintaining the myths about its waste products as long as it creates jobs. Endless studies have been done, full of unrealistic scenarios about what might happen to nuclear waste. After all, as we saw in Chapter 8, the industry likes to keep itself busy. In *Why Nuclear Power Has Been a Flop*, Jack Devanney concludes, 'These people have no interest in seeing the used fuel problem disappear even if it means convincing the public that nuclear power is perilous.'[17]

Another critic, energy analyst Rauli Partanen of Finland, wonders what message the industry is sending to the public when it 'packs the waste in several layers of different protective materials, digs a hole that is half-a-kilometre [1,600ft] deep for the waste and fills it with concrete afterwards'. Partanen provides the answer: 'Who in their right mind would believe that the waste is anything but the most dangerous and deadly stuff on the planet?'[18]

So *of course* people protest when experts and authorities identify a good disposal site for nuclear waste. Local residents' concerns about the risks may not be based on facts, but they're confirmed by an impression that's nurtured by the nuclear industry and its regulators.

Remarkably often, the choice of disposal sites falls on sparsely populated places where politically vulnerable groups live, such as Native Americans at Yucca Mountain or Aboriginal peoples of Australia. This choice emphasises the impression that something sinister is going on. You can't convince people that nuclear waste is free of hazards if you want to hide it somewhere out of sight.

The fear of nuclear waste is an example of selective fear. Just as the atomic bomb crosses a moral boundary that other deadly weapons apparently do not, and just as any accident in a nuclear plant seems more serious than in any other industry, so nuclear waste is seen as a much bigger problem than any other waste.

The highly radioactive waste from nuclear plants is special, indeed, but in a good way. This is not just because there's so little of it. It's waste that does not leak into the air, soil or water, and has never killed or sickened anyone.

Yes, but... (part 1): will it really stay safe in the ground?
Stewart Brand's proposal for the temporary storage of nuclear waste clashes with the prevailing view in politics, science and the nuclear industry. We must take into account, it seems, that society could be seriously disrupted sometime in the next million years, so the spent fuel had better be hidden away where nobody can access it. Man is fickle, whereas nature is

not. Earth's processes are incomprehensibly slow, spun out over hundreds of thousands of years. As Belgian geologist Manuel Sintubin says: 'From the Earth's perspective, one million years is just a blink of an eye.'[19]

Geologists are quite good at establishing how the subsurface has held up. The Finns found that for the past 500 million years there was hardly any movement in the granite bedrock beneath the island where Onkalo is created. Geologists can also predict reasonably well which subsoil will remain stable over the next thousand millennia. It's a whole lot easier than predicting what the climate will be like in 2050 or the outcome of upcoming elections.

Once the nuclear resources have been handed over to the Earth, we'll have nothing to worry about. Radioactivity will seep out eventually, but there will be no noticeable effect on the Earth's surface. Sintubin says, 'Even if humans went extinct, the biosphere will never be affected.'

There's no lack of suitable places for underground storage. Salt deposits like those in New Mexico, granite as in Olkiluoto or clay layers like the Boom Clay in the Netherlands and Belgium are being considered as ideal options. Such places are plentiful. In Germany, 90 areas were found suitable, covering more than half of the country's land area.[20] A panel of geologists in Denmark indicated that almost the entire country would be fine to store waste.[21]

Indeed, nature knows how to deal with nuclear waste. That's what happened in Oklo, Gabon. In 1972, French geologists discovered something strange at a uranium mine. The mix of isotopes indicated that the uranium had already undergone nuclear fission. What was going on here? Research revealed that a series of natural nuclear reactions had taken place at Oklo almost two billion years ago. Around a rich uranium deposit and flowing groundwater, Mother Earth had spent hundreds of thousands of years building a cluster of underground nuclear reactors, generating and sustaining fission, and letting the whole thing burn out on its own.

Today, the porous sandstone at Oklo would not qualify for final disposal, but even this subsurface was stable enough to ensure that in those two billion years, the naturally produced 'nuclear waste' has moved at most a few metres.

Yes, but… (part 2): what if things do go wrong thousands of years from now?

The Finns have gone to great lengths to consider what could go wrong.[22] Suppose the copper canisters were damaged unnoticed at the outset. Suppose the canisters as well as the bentonite clay inexplicably dissolved after only 1,000 years. Suppose something unexpectedly shifts in those deep geological layers and the groundwater moved upwards. Then suppose in 10,000 years someone lives in perfect ignorance on top of the most contaminated spot, grows all his food on that few square metres and gets all his water from a local spring. They truly put all this into the computer. And? How much radiation would that unfortunate descendant be exposed to?

Just 0.00018 millisieverts. That's the same amount you get when you eat a bunch of bananas.

Environmental groups immediately complained that the study was biased, because it was conducted by experts commissioned by Posiva, the company building the repository. Yet even the authors of a study commissioned by Greenpeace had to conclude that 'there is no proof so far that the planned repository is not safe'.[23] Presumably they couldn't think of a scenario bizarre enough to result in a safety problem.

Yes, but… (part 3): might something go wrong while hauling nuclear waste to a final repository?

Every year, countless packages of radioactive material are transported by road, water or rail. They are transported in such a way that in the event of any imaginable misfortune the material will remain undamaged. Never in the history of nuclear waste transport has there been an accident where radioactivity was released causing harm to humans or the environment.

At one time, things went awry. On 7 November 2004, a train transported 12 containers of nuclear waste from the La Hague reprocessing plant to a salt mine in Germany. Sébastien Briat, a 21-year-old activist from France, chained himself to the track. He ignored sabotage instructions and positioned himself after a bend, in a forested area, with no one else present to warn the driver with smoke signals or flags. The train was unable to stop and Briat lost his leg. On the way to hospital, he died of his injuries.

In the media, campaigners pointed out that his death illustrated the dangers when nuclear waste is transported.[24]

Yes, but… (part 4, final): nobody wants to live next to nuclear waste, right?

Not so fast. In Sweden, a site for final disposal was found with the support of local residents. During the process to find a location, municipalities could apply if they were interested in hosting the nuclear waste and could demonstrate local support. This led several municipalities to compete with each other. Talking to residents, businesses and experts, they discussed the advantages (employment, economic growth) and disadvantages (construction inconvenience, possible loss of property value). At all times, the population could request information and speak out against the plans.

In January 2022, the Swedish government agreed to designate the location as Forsmark, only 100km (60 miles) away from Stockholm. At the press conference, the environment minister said: 'Our generation must take responsibility for nuclear waste.'[25] At Forsmark, at least 1,500 jobs will be created immediately. The model around open and voluntary decision-making was also applied in Finland. Countries like Canada, Japan and the UK are following suit.[26]

It helps if a municipality already has a nuclear plant nearby. People are used to it. Some of them work there or know someone who does. It's not without reason that Onkalo is being built near the Olkiluoto nuclear plant and the Swedish repository near the Forsmark one. People living near a nuclear plant consider nuclear energy the most normal thing in the world and have confidence in the technology. Local cooperation, not the expertise of geologists, is likely to become the deciding factor in the answer to where the nuclear waste will go.

However, the question remains why each country has to come up with an individual solution. Already, the Swedes are building something similar to the Finns, while Onkalo has plenty of room for expansion. The French are building their own repository in clay. The Canadians are close to deciding on their approach.

The costs of deep geological disposal will come down with international cooperation. It's not inconceivable that countries will come forward to make money by providing a repository for a fee.

Everyone seems to love Onkalo. In international media, the repository has been widely praised. The *New York Times* stated: 'On Nuclear Waste, Finland Shows U.S. How It Can Be Done'.[27] Another headline, elsewhere: 'Nuclear power is indeed a good alternative to fossil fuels, now that the radioactive waste problem has been solved'.[28] The International Atomic Energy Agency calls Onkalo a 'game changer'.[29] Because now that the waste issue has been settled, nuclear energy has, at last, a future.

Still, the question remains how exactly Onkalo changes the game. It may well be that the geological repository doesn't lead to a revival or even an acceptance of nuclear power, but rather to its inevitable end. After all, now that we know what the final destiny of its legacy looks like, we might as well get rid of it.

Already, geologist Manuel Sintubin has argued that final disposal is 'part of ending the nuclear age'. He believes it's the only solution if we wish to 'put an end to it once and for all'.[30]

The final disposal of the waste from our nuclear plants has the symbolism of a farewell ritual. We wrap it up and bury it.

In the crypt below, far away from our civilisation, it stays out of sight for ever, with only a plaque and a message for posterity. This is what we do with the dead.

Onkalo will be shut down completely, just as the Egyptians did when they buried their pharaohs in pyramids, which were also built for eternity. It looks as if Onkalo will become the burial ground not so much of a bit of waste, but of an entire technology.

What exactly do we bury in Onkalo? We bury the monster we birthed. Ever since nuclear fission showed itself to the world in a bomb, we have turned away. *Nuclear power? No, thanks!* We don't even know what nuclear waste looks like; we don't *want* to know, because we have no real interest. Meanwhile, we ascribe mystical, horrific powers to it, as if it wants to escape and take revenge on its creator, the human species found guilty of hubris.

We fear nuclear waste because we consider it a curse. We see it as the debris of a sinful technology. We've gone too far, we've tinkered with nature too much. And so the monster must go away, somewhere in the caverns of the Earth, tucked deep down, where no one can ever kiss it awake.

Onkalo brings us redemption.

Nuclear waste can be stored safely. This is technically, economically and politically possible. But: is the deep geological repository really a good idea? Aren't we just creating this place because it makes us feel better? As James Lovelock, renowned creator of the Gaia theory, once said of the repository at Yucca Mountain: 'We need it about as much as we need a facility for imprisoning dangerous extraterrestrials.'[31]

A new movement of 'dark ecologists' doesn't want to hide nuclear waste away. They want to make the detritus of modern society visible, so that it becomes part of our living environment. British philosopher Timothy Morton, author of books such as *Dark Ecology* and *Being Ecological*, once suggested putting nuclear waste 'in small quantities in market squares, in a large glass shelter'.[32] That way we would be permanently reminded of it while keeping a closer eye. We would build a bond with the waste.

Morton draws on an old spiritual principle: we should not run away from our demons, but acknowledge and embrace them. He classifies nuclear waste among so-called 'hyperobjects', which he sees as 'the demonic inversion of the sacred substances of religion'.[33] Dark ecologists like Morton advocate that we should not turn away from our creatures, but take responsibility for them.

Morton might have taken inspiration from *Frankenstein*. In Mary Shelley's 1818 novel, a chemist named Victor Frankenstein fashioned a friend for himself. In the beginning, this nameless creature was not unfriendly and even smiled. But Frankenstein was repulsed by his work's hideous appearance. He abandoned him. That's how the creature became a monster. French sociologist Bruno Latour has argued that the story of Frankenstein teaches us that we, as a creative species, should not turn away from our inventions, but rather care for them.[34]

How can this be done with nuclear waste? We can give it a second life. What is called 'spent fuel' isn't entirely spent, it's merely 'used'. The fuel rods coming out of the reactor still contain so much energy that we could put them back in at the front end of a nuclear plant. This is because a typical reactor is fairly inefficient; a staggering 95 per cent of the energy inside the uranium and plutonium remains unused. Such squandering was never seen as a problem as uranium is plentiful and cheap.

But as the role of nuclear power might increase, it's better to avoid mining and to spare nature if there's an alternative. The alternative is there. It's called 'nuclear waste'.

In a fast breeder reactor, the waste can be recycled. The closed fuel cycle makes sure that not 1 per cent but at least 99 per cent of the uranium is converted into energy. According to one estimate, the existing stockpiles of nuclear waste in Europe are enough to provide zero-carbon electricity for the whole continent for at least 600 years.[35]

Such reactors are nothing new. The breeder reactor which can recycle its own waste has existed since the early 1950s. But once uranium turned out to be abundant, it was much easier and cheaper to extract it from mines. The breeder reactor ended up sidelined. Much later, countries such as the United States, France and Japan would build prototypes. These projects all ended prematurely due to politics and economics. The industry stood by and watched, and everything remained the same.

It's a shame, really. The famous Brundtland Commission, which put the concept of sustainable development on the map in 1987, explicitly included the breeder reactor among the 'renewable sources', along with wood, plants and dung.[36]

At the moment, only Russia has fast breeder reactors in use. The BN-800, which has been supplying electricity since 2016 at a nuclear plant in Zarechny near Yekaterinburg, can run on highly radioactive waste. At the end of the process, only a few isotopes remain, some of which can be used in medicine and industry.

Designs for nuclear plants that use waste as fuel are on the drawing boards of many start-ups, including Bill Gates' TerraPower, as well as established companies such as Hitachi and Westinghouse. These reactors are waiting for investors, while investors are waiting for politicians.

Why? Everything surrounding nuclear power is tied up in complex regulations, which are even more complicated when it comes to waste management. Regulators are just not set up for recycling. The current treatment of nuclear fuel rods in a reprocessing plant for one-time reuse is permitted, but is still far from the closed fuel cycle in a breeder reactor.

If we want to find a solution to nuclear waste, if we want to produce lots of zero-carbon energy and if we want to spare nature, politicians will need to encourage the development of fast breeder reactors so that the nuclear industry can change accordingly.

It seems like the decision makers in Europe acknowledge this. Fast breeder reactors fit neatly in the plans for a circular economy, in which few materials are used and the waste gets reused. The 'circular economy action plan', launched in 2020, is part of the European Green Deal, a set of policies to reduce pressure on natural resources and achieve climate neutrality by 2050.[37]

These reactors can easily respond to the intermittent production of wind and solar; the combination makes Europe's climate targets feasible. It would be a small step to make recycling of nuclear waste in a fast breeder reactor mandatory for new nuclear plants.

Building so many reactors will surely cost a lot of money. But there's good news. For decades, the industry has collected pennies from each energy bill to save for the final disposal of its waste. This means there's now a considerable piggybank. Removing or simplifying the plans for geological repositories means tens of billions of euros will be freed up in Europe to start building a fleet of fast breeders. Such construction is radically different from Onkalo. It doesn't mean a *burial*, but a *rebirth*.

Long live nuclear waste!

No doubt future generations will look back on this time and wonder what on earth we were doing. As many seem to believe, they might ponder why we burdened them with nuclear waste. But will they really? Or will they rather blame us for having saddled them with all that other waste that is much less discussed? Maybe they will wonder why we burdened them with climate change even though there was a perfectly good alternative to fossil fuels available.

It could also be that they will be wondering something completely different when they learn that we buried barely used nuclear fuel deep underground: *why did they hide away such a valuable source of clean energy?*

CHAPTER ELEVEN

Dreaming of Progress

What does the future of nuclear energy look like?

'Perfect is the enemy of good.'

– Aphorism

It's a summer day in August 2017. Behind the quiet sand dunes near Petten, a village in the Netherlands where the North Sea is kept at bay, the only noise comes from a group of seagulls. They're fighting over the remains of a lunch someone has left behind. Other than that, the place is deserted. You'd be surprised to find that some hold this as a historic place and a historic day.

Here in Petten, at the facility of the NRG (Nuclear Research & Consultancy Group), a project has just begun that could be a milestone in the development of nuclear power. As of this morning, the research reactor contains six lumps of salt, each the size of a die and carrying a few grains of thorium.

It may not sound that spectacular. Yet thorium has been heralded as an improvement in nuclear technology, and could be the next step in the transition to a clean energy supply.

Thorium is a weakly radioactive element, which is even more abundant in the Earth's crust than uranium. A so-called molten salt reactor using thorium produces waste that's relatively short-lived, 'only' 300 years or so. The reactor is supposed to be entirely safe thanks to features such as a plug of solidified salt, which melts automatically when the reactor gets too hot, for instance if the cooling pumps lose power. Once the sealing plug has melted, the salt flows into storage vessels underneath the reactor and the nuclear fission process stops. Put simply: since the fuel is already molten, nobody needs to worry about a meltdown. If the reactor overheats, things will be fine, not thanks to human intervention but to the laws of nature.

Over a cup of coffee in the canteen, Sander de Groot, a product developer at NRG, can't stop talking about thorium. The first time he read about it must have been around 2006, he says. A molten salt reactor containing thorium had been running at the Oak Ridge National Laboratory in the 1960s. It worked. Everything that went on with the reactor was meticulously recorded at the time. For decades, these reports had been gathering dust in a filing cabinet. At the beginning of this century, a young employee of NASA rediscovered the reports, scanned them and made them available online. De Groot stumbled upon them a few years later.

Whenever De Groot talked about thorium with colleagues, they found it 'too wild' and 'speculative', he says. A smile appears on his face. 'Well, today, at last it's happening.'[1]

A lot has changed since – and not just for Sander de Groot, who moved on in 2022 to co-found Thorizon, a start-up designed to accelerate developments in the use of thorium. First there were the number-crunchers, checking to see if such a reactor could operate without any problems. Then came the entrepreneurs, looking for investors to support their business cases. And, now, it's time for the next step: fundamental research, financed with public money and carried out in a reactor owned by the European Union. The Netherlands is leading the way. Delft University of Technology is coordinating an international collaboration to develop the design of a thorium molten salt reactor.

'The potential is huge', says De Groot. 'In a molten salt reactor, thorium can be used so efficiently that it can be a safe and almost infinite source of energy.'

After decades of stagnation, a wave of innovation is sweeping through the nuclear industry. Thorium is one of many options for the future. Work is being done on fuels that can withstand every conceivable accident, and on reactors that can more easily cope with the erratic, weather-dependent power production of wind turbines and solar panels. Much of the focus is on small modular reactors (SMRs), for which components are produced in factories and assembled on site.

Industry representatives say new nuclear plants are becoming more efficient, requiring less uranium and producing less waste. Or they are becoming safer, thanks to yet another layer of measures. Some designs are intended for sites near coastal cities, others for remote areas. Some use both the power and the heat released to produce hydrogen, which can be used in industry or as fuel for aircraft and container ships.

And then there's nuclear fusion. In a fusion reactor, temperature and pressure are extremely high so atoms squeeze together like they do in stars, releasing an awful lot of energy. So far, nuclear fusion has always been something that exists only on the horizon. In 2021, a milestone was reached at the Oxfordshire-based Culham Centre for Fusion Energy where researchers managed to sustain fusion for five seconds, releasing 59 megajoules of energy, enough to boil a few dozen kettles of water. It's also enough to get dozens of headlines published across the world.

All these stories feel like a breath of fresh air. Could the nuclear industry finally be changing? Is there room for new and bold ideas that give nuclear energy a future? Might it be, after its dark journey through Hiroshima and Chernobyl, on its way to an awakening?

In the emerging pro-nuclear movement, there's plenty of discussion about the benefits and downsides of each new reactor design. There's a catch to everything. In one design, the material would be prone to corrosion, while in another, disposal of fission products may become a concern. Breeder reactors could dispose of nuclear waste, but they might also be used for the production of weapons material.

With their technical disputes over the pros and cons of each reactor, proponents of nuclear power inadvertently reinforce the main point of their opponents: *that today's nuclear plants are flawed*. 'Advanced nuclear', the generic term for all innovations, suggests that current nuclear plants are not advanced. It's as if they're obsolete and in need of serious adaptations.

And why is there so much focus on innovation, really? As we've seen in this book, nuclear plants have a pretty good track record. The problems with nuclear power have nothing to do with technology. The problems are between the ears.

Nuclear plants are distrusted, so now, with each innovation, they need to be made socially acceptable. Safer, because nuclear power is fraught with dangers. Flexible, because solar and wind should be prioritised. Smaller, because small is beautiful. But will all this innovation really make nuclear power better? Does anyone believe that a different type of reactor or fuel will make resistance disappear?

Business magnate Bill Gates invests in the start-up TerraPower, which wants to build its own nuclear reactor, and insists that the nuclear industry really needs to 'innovate and adapt'.[2] But while he has been pouring part of his fortune into this company since 2006, TerraPower has yet to build anything. Meanwhile, its design looks more and more like that of an experimental breeder reactor from the early 1950s, before Gates was even born.

Some say Bill Gates' involvement in advanced nuclear is making the technology more popular and more inviting. So far, his name has guaranteed media coverage about the need for nuclear power in the future. But Gates and other innovators aren't universally appreciated. Michael Shellenberger, a California-based opinion leader who did much to build the pro-nuclear environmental movement, believes Gates and his peers' main goals are social and political power. In promising a 'magic box' that is 'apocalypse-free', Shellenberger argues, they're creating a religious status for themselves. 'It's totally dishonest, it's absolutely manipulative of emotions, of people's good will,' Shellenberger said, speaking as a guest on the Decouple podcast. 'But nonetheless, it's the dominant strategy of pro-nuclear people.'[3]

In the PowerPoint presentations, any new nuclear reactor looks pretty, free of troubles. Westinghouse's AP1000, for example. This reactor was to become the showpiece of the US nuclear industry, which was once again keen to be at the forefront of development. In 2013, construction began on four such reactors. However, at the Vogtle plant in Georgia, it took more than a decade to connect two AP1000s to the grid. Construction of the other two, in South Carolina, was cancelled.

Or take the innovative European reactor known as EPR. Construction began in 2007 in Flamanville, Normandy, and it's now scheduled to open in 2024. Meanwhile, the Finns opened their EPR in 2023. It took them 17 years to build, 12 years longer than planned and several billion euros more expensive.

In the early 1950s, Hyman Rickover was an admiral in the US Navy who built a nuclear submarine with a small pressurised water reactor on board, which today is the basis for nine out of every ten nuclear plants around the world. When it comes to new reactors, this quote by Rickover from 1957 is one to remember:

> *Any plant you haven't built yet is always more efficient than the one you have built. That is obvious. They are all efficient when you haven't done anything on them. They are in the talking stage. Then they are all efficient. They are all cheap. They are all easy to build, and none have any problems. That is quite correct. They do not have any problems at that stage.*[4]

Technical discussions about the next step in nuclear power seem to lead, most of all, to procrastination. If you really want nuclear, only one thing works: start building. Build the nuclear plants that have proved themselves. Those who opt for existing designs benefit from the experience of constructors, engineers, operators and regulators. Standardisation and serial construction are essential to get nuclear plants ready faster and cheaper. A willing and unwavering government can enforce this. It's as simple as that.

The choice may well be the AP1000 or the EPR. While construction was delayed in Europe and the US largely due to political and organisational problems between the nuclear industry, regulatory bodies and the government, China didn't have those issues and built its first two EPRs in less than a decade. Larger orders will speed things up. Ukraine has ordered nine AP1000 reactors. France wants to add a good number of EPRs.

This is how things start moving. This is how nuclear power might have a future.

As a child, Princy Mthombeni got up at 4 a.m. She walked for half an hour in the dark to fetch water. Back home, before walking to school, she got down to preparing breakfast. For cooking and heating, the Mthombenis used paraffin, which was expensive, so they also scraped together charcoal, branches, twigs and leaves, supplemented with dried cow dung. It kept Princy and her two live-in cousins busy. She had to finish her homework before sunset; her village, Nqutu in KwaZulu-Natal, South Africa, had no electricity and there were few candles at home. When evening fell, the only light came from the stars and the moon.

Princy Mthombeni's story will be familiar to some three billion people across the world. It's estimated that around a billion lack electricity; the rest have only a little. Things are often better in cities than in villages, but a power outage can last for weeks. Africa performs worst in providing reliable electricity.[5]

As a teenager, Mthombeni moved to the city and a house with electricity. Her mother explained that the light bulb on the ceiling flashes on when you flip a switch on the wall. *Magic!* They used the electric cooker to prepare food for up to five days, keeping the extras in the fridge. 'Electricity made our lives much easier.'[6]

Later, Mthombeni went to live with her uncle and his wife in a township in Johannesburg where she could continue her education. She had a knack for maths and took a computer course. This is how she learned to quantify energy poverty in Africa. Citing figures from the Africa Progress Panel, chaired by former UN Secretary-General Kofi Annan, Mthombeni sums up: 'A freezer in the United States consumes ten times more electricity than a resident of Liberia uses in one year. A kettle boiled twice a day by a British family uses five times as much electricity as a person in Mali uses per year.'[7]

Lack of reliable energy puts a brake on progress. Many people, young and old, work in the fields. Women have to cook and do manual washing every day. In hospitals and health clinics, lights fail during surgery or childbirth, and without refrigeration, medicines and vaccines lose their shelf life. Electricity in particular is crucial for development and emancipation, experts know. Modernisation comes with the socket.

Billions of people are expected to consume more energy – a lot more energy, especially electricity. By 2050, global energy demand, which includes heat and transport, will increase by half compared to 2020. Electricity demand is growing even faster. Fifty per cent growth is plausible before 2040. Africa and Asia will account for most of it.[8]

What will they do with all that electricity? They might buy fridges and store food. They might buy televisions, light bulbs for every room, computers, washing machines, stoves. And once they can afford it, they might buy fans to provide some relief from the scorching heat. According to the International Energy Agency, sales of air conditioners will skyrocket in the coming decades. Over the next 30 years, 10 units will be sold every second.[9]

How will all this energy be generated? Over the past 25 years, countries like China, India and Brazil gave the answer: by burning fossil fuels. The resulting economic growth in these countries is the main reason for the increase in CO_2

emissions since the 1997 Kyoto Protocol. Coal costs a pittance, there's a lot of it and a plant is built in no time. Infrastructure for a gas plant is a bit more complex, but after a few recent discoveries of natural gas fields, there's plenty to burn for many more decades.

If poor and upcoming countries want to build a coal or gas plant, they can no longer obtain a loan from development banks like those of the European Union or even the World Bank, which says it is 'committed to fighting poverty'.[10] Nor can they turn to wealthy countries gathered in the G7 group, including the United States, the United Kingdom and Germany.[11] At the Glasgow climate summit in November 2021, more government leaders pledged to stop funding foreign fossil fuel projects.[12] Their decision marked 'a significant boost' for the energy transition, cheered the *Guardian*.[13]

Elsewhere, the pledge drew criticism. Vijaya Ramachandran, an economist who worked for the World Bank and is affiliated with the Center for Global Development, called it 'hypocrisy'.[14] In an opinion piece in *Foreign Policy* magazine, she pointed out that while these government leaders are putting the brakes on fossil fuel use in poor countries, they're not doing so within their own borders. Just one month before signing the pledge at the climate summit, Norway's prime minister fervently defended the continued drilling of its large natural gas reserves as part of a 'transition that is needed to succeed in the momentum toward net zero'.[15]

Ramachandran speaks of 'green colonialism'. What else to call it when rich countries, in the name of protecting the Earth, tell poor countries what they should and should not do?

Samir Saran, an opinion leader in India, concluded: 'Perpetuation of global poverty and low incomes cannot be the rich world's climate mitigation strategy.'[16]

The promise to stop funding fossil fuel projects in poor countries has nothing to do with climate justice. Historically, Africa's contribution to global warming has been negligible. By 2020, the CO_2 emissions of over a billion people in sub-Saharan Africa totalled no more than 3 per cent of all global greenhouse gases.[17]

Diverting funding from fossil fuels frees up billions of dollars a year to support wind turbines and solar panels. Princy Mthombeni thinks this will be insufficient. After all, with intermittent power, industry in Lagos or Lusaka cannot take off. 'There is a misplaced idea held by some people within the global community that Africa's relative lack of legacy infrastructure makes it the perfect canvas on which to paint a green energy future,' Mthombeni believes. 'They're putting restrictions on the energy choices of poor countries which are continuously trapped in a state of energy poverty, socioeconomic challenges and underdevelopment. But people don't care where their electricity comes from. They just want to be able to turn on lights and microwave their food.'

Mthombeni chooses nuclear power. She's the founder of Africa4Nuclear, an action group that makes the case for nuclear to power the future of the poorest continent.

Among the dozens of countries making plans for nuclear plants, Africa is strongly represented: from Algeria and Sudan in the north to Namibia and Zambia in the south, and from Senegal and Ghana in the west to Kenya and Ethiopia in the east.[18] Currently, most of these interested countries get their energy from a combination of fossil fuels and hydropower.

These plans are becoming more concrete. Intentions are being worked out and agreements are being signed with companies, mainly from Russia and China, but also from South Korea. These firms will have to provide the technology, train personnel and set up regulatory bodies.

Currently, Africa has only one nuclear plant, located north of Cape Town, which opened in 1984. Before an African nuclear fleet becomes reality, there's still a long way to go. Construction requires high-risk investment. Political instability and corruption create weak institutions, making financing difficult and sabotage possible. Then there are the technical obstacles. For instance, in many African countries, the electricity grid is not designed to handle the large power output of even a single nuclear reactor, which in the West could easily supply enough for a million households. Cooling requires a lot of water, which is not always available. Safety systems require the constant attention of well-trained staff. Before construction starts, a whole list of conditions must be met. For now, Africa does not seem ready for nuclear power.

Fortunately, there's innovation! Advanced designs offer a solution to all problems, or so it seems in the industry brochures. New reactors are small, flexible, super-safe and cheap. They're easy to build, simple to operate and can be placed anywhere. Who wouldn't want such nuclear plants?

There's just one drawback: they don't yet exist.

Princy Mthombeni, however, remains confident. 'Nuclear power is clean, reliable, dispatchable and baseload while mitigating against climate change,' she insists. With those qualities, nuclear power is in line with the international community's aspirations for sustainable development.

All this was confirmed in a 2021 report by the United Nations Economic Commission for Europe (UNECE). One of the conclusions: nuclear energy is an 'indispensable tool' to achieve sustainable development goals. 'It has a crucial role in providing affordable energy and climate change mitigation, as well as eliminating poverty, achieving zero hunger, providing clean water, economic growth, and industry innovation.'[19]

To get there, the West may need to offer some help. Without it, African governments will opt for coal and natural gas, supplemented by solar and wind.

Electricity demand will increase not only in poor and emerging countries, but also in prosperous ones. This is largely because we want to electrify as many aspects of our lives and industries as possible. Experts also foresee the further growth of power-hungry enterprises such as robotics, biotechnology, data centres and hydrogen production.

Add the failure of climate policies, and it's not surprising that interest in nuclear power is rising among major economies such as the United States, Japan and China. There's growing awareness that solar and wind cannot out-compete fossil fuels; the real competitors are uranium atoms.

Interest has also been building since the return of geopolitical tensions between the West and fossil fuel-rich Russia – and, who knows, one day perhaps with China, the undisputed market leader as a supplier of rare Earth metals such as lithium, manganese, gallium and neodymium, which are needed in wind turbines, solar panels and electric car batteries. After the Russian invasion of Ukraine, the Belgian government suspended the planned closure of two nuclear reactors by 10 years.

By any measure, uranium is plentiful. Put a shovel in the ground, anywhere, and you dig up a bit of uranium from just about any pile of earth. As a result, fuel for nuclear plants is hardly susceptible to political conflict. Uranium mines are found in such geographically diverse countries as Canada, Australia, Kazakhstan and Namibia. There's much more uranium to be found in the oceans, where the supply is renewable: when you remove uranium, it replenishes itself, thanks to the interaction between seawater and rock.

If all construction plans proceed – in African countries interested in nuclear power, in Asia where China and India are leading by example, in European countries resisting pressure from Germany – the capacity of all nuclear plants worldwide will increase by 26 per cent by 2050, as reported in the 2021 edition of the *World Energy Outlook*.[20]

It may sound like an impressive growth rate. However, electricity demand is growing much faster. If all plans proceed, the *share* of nuclear power in the global energy mix will continue to shrink.

Popular support for nuclear power is on the rise, too. Surveys in Western countries show that more people are in favour of nuclear power than against it. And the support is growing fast, as shown by established opinion polling agencies.

In the US, a 2023 survey showed 57 per cent want more nuclear plants, up from 43 per cent in 2020.[21] During the same period, support for nuclear in the Netherlands grew by half, with twice as many proponents now as opponents.[22] In Poland, support grew even faster: from 39 to 75 per cent in only one and a half years, largely due to the desire to gain energy independence from nearby Russia.[23] In Belgium, 85 per cent want to keep nuclear in the energy mix with investments in SMRs.[24] In Sweden, 84 per cent want to continue to use nuclear power or build more reactors. Only one in ten respondents want to shut down nuclear plants.[25]

This may come as a surprise. Open a newspaper or turn on the television, and any item on nuclear power suggests that nobody wants it. Journalists who wouldn't think of quoting anti-vaxxers as experts in an item on health care have no qualms in quoting anti-nuclear activists as if they were reliable experts on energy policy or environmental issues. It's a stark contrast to how journalists portray activists against, say, wind turbines.

The public support for nuclear need not be wholehearted or underpinned by technical knowledge. Perhaps that support should be qualified as tolerance. It could be fragile; a nuclear accident, anywhere, could change views just like that. But still, the support is there, despite the negative portrayals in the media. There is, however, one important difference: opponents are more strongly against nuclear than proponents are in favour of nuclear.

In politics, something else is apparent: supporters are mostly found to the right of centre. In parliaments, outspoken support for nuclear power often seems like bullying to those on the left. It leads to uneasy alliances. Just as supporters of solar and wind unwittingly make a Faustian bargain with the fossil fuel industry to supply back-up power (see Chapter 9), climate activists for nuclear power unwittingly link themselves to a political colour that is not necessarily theirs. In fact, many of them are annoyed by the relaxed tone on the right when it comes to climate change or the environment.

However, nuclear energy does offer something for everyone. Progressive thinkers value science and a knowledge-based economy. Socialists like to see low energy bills and solid jobs. Conservatives welcome an energy source that doesn't require imports from authoritarian regimes. Conservationists would rather see a compact clean energy source somewhere on an industrial site than scattered all over the landscape in fields and meadows. This makes nuclear one of the rare climate policies that can attract support from both sides of the aisle.

Interestingly, when the newly elected centre-right government in Sweden announced in 2023 that it was planning a 'massive build out' of its nuclear fleet to drive the electrification of industry and transport,[26] one newspaper framed it as part of a climate policy that was 'less ambitious' than the one pursued by the previous, centre-left government. 'The right-wing bloc,' the article stated, 'is no longer aiming for a fully renewable energy supply, but for a 100 per cent fossil-free future.'[27]

Apparently, to some, it's ambitious to aim for all-renewables, but not to aim for all-non-fossil fuels.

While backing for nuclear can be found in the constituency of every political party, including Social Democrats and Greens, at the top of these parties the support is typically lacking. Why is that? One possible answer lies in the background of politicians. Among members of parliament, there's a strong over-representation of those trained in humanities or social sciences such as history, sociology and communication studies. Very few have knowledge of things like physics, engineering or computer science. This is a shame, as those with a background in natural sciences often bring a rational and analytical view to political issues. Diversity in a group tends to lead to better decisions.

Many supporters of nuclear power believe they are a well-informed minority standing up to an immense army of ignorant opponents. In reality, they represent a broad majority in society while the opponents consist of a small but influential legion of dogged activists and financial stakeholders in the establishment, supported by naive dreamers in politics.

By now, near the end of this book, such observations should come as no surprise. With nuclear power, hardly anything is as it seems.

Changing minds is never easy. Warning about climate catastrophe in 2004, James Lovelock, an environmentalist's hero, made an impassioned appeal in a piece in the *Independent*: 'I entreat my friends in the movement to drop their wrongheaded objection to nuclear energy.'[28]

George Monbiot followed this advice. Back in 2000, Monbiot wrote in the *Guardian* that nuclear power was 'the world's most dangerous business', arguing it was 'time to shut nuclear power down'.[29] When Monbiot learned the accident at Fukushima in 2011 wasn't nearly as bad as the media made it out to be, he became convinced of the benefits of nuclear power (see Chapter 6). He wrote a piece in the same newspaper explaining how he learned to 'love' nuclear power.[30]

But Monbiot faced a lot of criticism, and would soon come to regret his change of heart. In 2013, he clarified his position by saying he did not support nuclear power 'at any cost'. For the EPR that was going to be built at Hinkley Point, Monbiot considered the price 'too high'. Besides, that reactor design was 'clunky', he continued, as it 'already looks outdated, beside the promise of integral fast reactors and liquid fluoride thorium reactors'.[31]

Two years later, Monbiot reflected on his conversion to the cause of nuclear power. Writing in the *Guardian*, he called it 'painful and disorienting', as his new position was 'antagonising friends and alienating colleagues'. Perhaps Monbiot was trying to win them back. Upon confessing that he still believes nuclear power has 'great potential to balance the output from renewables', he called on the government to 'kill' the planned reactor at Hinkley Point. Instead, the money should be spent on solar panels and wind turbines and on 'a comparative study of nuclear technologies, including the many proposed designs for small modular reactors'.[32] Procrastination, anyone?

It looked like George Monbiot had discovered the therapeutic value of innovation. Nuclear power as we know it is haunted by accidents, a violent past and a lying and blundering industry. Today, hardly anybody seems to like nuclear, and in any case, all things nuclear take too long and cost too much. But advanced nuclear power, *i.e.* nuclear power that *does not yet exist*, who can be against that? Innovation offers an escape from the boring, depressing world of 'old' nuclear power.

There's another psychological advantage to dreaming of future nuclear technology. It allows long-time opponents to put aside their objections without losing face. This happened, among others, to Myriam Tonelotto, a filmmaker from Italy who protested against nuclear power and campaigned against the dumping of French radioactive waste off the coast of Somalia. While making a documentary about the work the European Commission was doing to develop nuclear technology, she was on her guard while talking to researchers. Then one of them started talking about thorium.

Tonelotto had never heard of thorium, and took to the internet to learn more. She read that a molten salt reactor using thorium was 'inherently safe': a phrase she had distrusted for years because the nuclear industry had been using it for a long time. But now it made sense to her. Little by little, Tonelotto learned that there might be a good kind of nuclear power after all. She made a documentary about it, *Thorium, la face gâchée du nucléaire* (*Thorium: The far side of nuclear*). 'If you think about it rationally,' says Tonelotto, 'the facts clearly speak in favour of thorium.'[33]

Do they, really? Are the 'facts' about a reactor that exists only in PowerPoint truly more convincing than those regarding the hundreds of nuclear reactors that have been in operation for decades? Such thinking leads to quotes like this one: 'I'm not against nuclear power, I'm against the technology currently in use.'[34]

That's not Myriam Tonelotto speaking. That's Wim Turkenburg – you know, the man who featured in Chapter 3 as a co-founder of the Dutch anti-nuclear movement and in Chapter 6 as a TV expert who was convinced that radiation would kill all workers in Fukushima.

Not everyone is circling the issue. 'Personally, I am against nuclear power,' Greta Thunberg wrote in a post on Facebook in 2019. She admits that the IPCC's climate experts consider nuclear a solution, 'even though it's extremely dangerous, expensive and time consuming'. Greta concludes: 'But let's save that debate until we start looking at the full picture.'[35]

What exactly is that full picture? Climate change, she says, 'is only a symptom of a much larger crisis'.[36] She says 'we have to change our way of life'.[37] This includes 'practically everything', including 'a whole new way of thinking'.[38] We need to 'change the system'.[39]

Such calls are heard ever more frequently. Naomi Klein, a figurehead on activism front lines since her 1999 antiglobalisation book *No Logo*, argues that climate change is essentially not about CO_2, but about capitalism. After all, what is needed to reduce greenhouse gas emissions would be 'extremely threatening' to an elite that holds much power over the economy, politics and mainstream media.[40] Capitalism, therefore, needs a reset. In her book *This Changes Everything: Capitalism vs the Climate*, Klein argues that we should consider the climate crisis as an invitation to transform our societies and build 'something radically better'.[41]

During the unprecedented economic growth spurt of the past 200 years, the natural world was degraded. If billions more people experience such growth, it could be disastrous for the planet as well as for them. Meanwhile, some companies have become so big that their value exceeds the GDP of entire countries. Over time, corporations have gained an unhealthy influence over politics, making decisions that too often revolve around money, not people. From a global perspective, the abject wealth of the few clashes with the abject poverty of the many. Indeed, something has failed, and things could be better.

That 'something radically better' Klein hints at will have to manage with a lot less energy. She is convinced that there are 'deep changes required' to our energy consumption.[42] Ideally, she believes, all of us would become energy producers, with solar panels on our roofs and maybe community-owned wind turbines outside town. This would provide citizens with democratic control over energy. With approval, Klein quotes the slogan of the environmental justice movement: 'System Change, Not Climate Change'.[43]

It might be radically better, indeed. But for the time being, system change has the practical disadvantage that it won't heat your home.

Greta Thunberg and Naomi Klein may be opinion leaders, but they have both paid an almost humble visit to a man who is an inspiration to them: Jorge Mario Bergoglio from Argentina. Once, he was a bouncer at a bar, later he swept the floor as a caretaker and worked as a technician in a laboratory. Then, Bergoglio gained world fame – at the age of 76 – when he was elected head of the Roman Catholic Church. Since then, everyone knows him as Pope Francis.

In good Christian tradition, the Pope has harboured dark thoughts about the future. The growing popularity of the Green movement seems to have inspired him to use new language for old ideas about nature's wrath. In an acclaimed environmental encyclical, *Laudato si'*, published in 2015, he wrote: 'Doomsday predictions can no longer be met with irony or disdain. We may well be leaving to coming generations debris, desolation and filth. The pace of consumption, waste and environmental change has so stretched the planet's capacity that our contemporary lifestyle, unsustainable as it is, can only precipitate catastrophes.'[44]

Naomi Klein, on visiting the Pope: 'This encyclical speaks to people's hearts.'[45]

Greta Thunberg, after shaking his hand: 'Thank you for speaking the truth.'[46]

How can we create a society that produces energy without disrupting the climate? That's a tricky question. Very few countries have accomplished this. Greenpeace found seven. In a 2018 video, they tell us that these countries run on 100 per cent renewable energy. We see upbeat animations of wind turbines and solar panels. People are happy, too. A couple in love take a selfie. A man waters flowers. A woman cycles by. Down the road is a vegetable stall.[47]

Where is this paradise?

Greenpeace doesn't mention the countries names, but we do get to see seven national flags. After some online digging, we learn that these nations barely use solar and wind, but run almost 100 per cent on hydropower, geothermal or biofuels. They include Lesotho, Bhutan, Nepal and Ethiopia.

Perhaps Greta Thunberg has seen the video. After all, she believes that we should turn to the poorest half of the world's population for inspiration, since this half accounts for only 10 per cent of total CO_2 emissions.[48]

In Lesotho, there's certainly little overconsumption. There's also hardly any economic growth. In fact, per capita income in 2020 was lower than it was a decade before, with two in three residents living below the poverty line. Life expectancy (under 55) ranks among the lowest in the world. So, can we say that the people of Lesotho discovered the right 'way of life' and developed the right 'way of thinking'?

Would the people of Lesotho recognise themselves in Greta's criticism of world leaders, whom she believes only talk of 'money and fairy tales of eternal economic growth', and accuses of having 'stolen my dreams and my childhood'?[49]

Princy Mthombeni suspects not. 'People here don't even know Greta. Personally, I find her words hypocritical. Let Greta come and live in an African village, even if only for a week. Would she still say we should leave energy sources in the ground? I'd like her to experience life here before she says such things.'

Of course, Greta Thunberg does not really want poor people to stay poor. She spoke those words as a teenager. In fact, she has shown herself to be well aware of global inequality. On the stage where she was invited to speak by the powers that be, she added: 'And yet I'm one of the lucky ones.'[50]

This was no lie. Her home country scores high in all rankings, including well-being, education, healthcare, democracy, emancipation, leisure time, life expectancy, economic equality, social cohesion, LGBTQ+ rights, happiness, clean air, animal welfare and free press. Her mother was an internationally celebrated opera singer who achieved her breakthrough at the Eurovision Song Contest when Greta was six years old. Before Greta went truant, she attended the same private school where Sweden's royal family was educated.

Greta Thunberg's desire for other people to change their habits and thoughts is reminiscent of the analyses made from the 1950s onwards by E.F. Schumacher, the author of *Small Is Beautiful* (see Chapter 3), who, as the son of a distinguished professor, went on to study in Berlin, Oxford and New York and worked as a trader on both the Wall Street and London stock exchanges before becoming a top adviser to the British coal industry.

Schumacher, too, saw no merit in having poorer countries use the modern technologies that had brought so much good – oh no, wait, so much *misery* – to countries that were now rich. He believed we could learn something from poor countries and liked to speak of 'the right path of development'.[51] He argued that people in impoverished nations should stay in their villages and insisted on 'appropriate' technologies,[52] which were cheap and labour-intensive and did no harm to the environment – unlike the coal industry, to which he owed his prosperity, allowing him the leisure to contemplate.

How different was the view of C.P. Snow, who at the same time (and in the same chapter) analysed the clash between scientific culture and intellectual culture. Snow, whose father worked in a shoe factory and earned extra money as a church organist, became the first from his school to win a scholarship to go on to higher education thanks to high grades. Later, he was reproachful towards his fellow scientists for not understanding anything about literature. And he was reproachful towards his fellow writers for not understanding anything about the Industrial Revolution. Intellectuals, Snow believed, were 'natural Luddites', opponents of technological development.[53] To Snow, industrialisation was 'the only hope of the poor':

> It is all very well for us, sitting pretty, to think that material standards of living don't matter all that much. It is all very well for one, as a personal choice, to reject industrialisation – do a modern Walden, if you like, and if you go without much food, see most of your children die in infancy, despise the comforts of literacy, accept twenty years off your own life, then I respect you for the strength of your aesthetic revulsion. But I don't respect you in the slightest if, even passively, you try to impose the same choice on others who are not free to choose.[54]

Further back in time, in 1901, French novelist Émile Zola, who grew up in a poor family and married a worker's daughter, wrote expectantly about the much-desired liberation of the fate of the lower classes: 'The day must come when electricity will belong to everybody, like the water of the rivers and the breezes of the heavens. It will be necessary to give it abundantly to one and all, and to allow men to dispose of it as they choose. It must circulate in our towns like the very blood of social life.'[55]

What a contrast to the words of Paul Morand, who studied at universities in Paris and Oxford, worked as a diplomat and married a princess who happened to be the daughter of a successful banker. Speaking of the same era, Morand scowled: 'Electricity is accumulated, condensed, transformed, bottled, drawn into filaments, rolled upon spools, then discharged under water, in fountains, or set free on the house tops or let loose among the trees; it is the scourge and the religion of 1900.'[56]

Could it be that from the position of a certain standard of living, the temptation lurks to renounce progress and glorify a life of greater simplicity?

Princy Mthombeni: 'We call it "privilege".'

We are, with apologies, going to bring up Greta Thunberg again. After all, she's the figurehead of a movement, someone with a widely supported narrative in the public debate. When she speaks, everyone listens. If Greta and so many others reject nuclear power, despite its obvious usefulness in fighting what they consider an urgent crisis, we should ask: are they really interested in a solution?

Greta herself does not seem so sure. She often places the word 'solutions' in quotation marks to convey sarcasm.[57] People who see climate change as merely a problem to be solved would deny, she claims, that climate change is in reality 'an indication of a system error'.[58] Once we reduce climate change to just a problem, someone might come up with a solution, and Greta insists we would still be stuck with that system error.

Her father agrees. Referring to his thoughts on nuclear power as a younger man, Svante Thunberg says on the final pages in *Our House Is on Fire*:

> 'I wanted to believe that people could fix everything. That we had managed to find all the solutions. Because if we had succeeded, we wouldn't need to change.'[59]

We're getting to the point. Apparently, it's not about the problem or the solution, it's about us: we need to change.

According to this perspective, we need to learn when enough is enough. There's no need to always want more, bigger, faster and better. We should, instead, develop a little more humility, modesty, sobriety. Let's not keep altering the natural world but instead take responsibility to care for it, appreciate it and accept its boundaries.

Writing in *This Changes Everything*, Naomi Klein even praises renewable energy sources for inviting us to 'pay attention to things like when the sun shines and when the wind blows'. She maintains that renewables 'require us to unlearn the myth that we are the masters of nature – the "God Species" – and embrace the fact that we are in relationship with the rest of the natural world.'[60]

Pope Francis couldn't agree more. In *Laudato si'*, the appointed Vicar of Christ wrote: 'The harmony between the Creator, humanity and creation as a whole was disrupted by our presuming to take the place of God.'[61] Surrounded by precious art treasures in his palatial residence in the Vatican, he continued: 'Humanity is called to recognise the need for changes of lifestyle, production and consumption.'[62] And: 'Many things have to change course, but it is we human beings above all who need to change.'[63]

We see a similar call in the statutes of Greenpeace. One of its official missions is 'to help bringing about a fundamental change in Man's way of thinking'.[64] Another objective reads: 'to end all nuclear threats'.[65]

There is a clear link between these two desires. Nuclear power gave – no, *gives* – us the ability to 'fix everything', to quote the actor Svante Thunberg. An end to wars as we have known them. An inexhaustible, clean source of cheap energy that can preserve nature. The ability to desalinate seawater so that there is enough drinking water. The ability to diagnose and treat diseases with radioactive isotopes. The possibility of a luxurious life like privileged people have.

Princy Mthombeni from South Africa nods. 'Yeah, that would be nice, indeed.'

'Fix'. Thunberg probably didn't mean that in a positive way when he said it ('I wanted to believe that people could fix everything'). According to people like him, the world's major problems must be solved by a change in our inner selves. We can end war through cultivating compassion and entering a dialogue. We must stop environmental pollution by tempering our ambitions. They stress that our connection with nature must be restored.

When an improvement emerges through a technical invention, it's dismissed by some as a 'technofix' – a denigration of the potential technology has to improve lives. It's what Anurag Saha Roy, a young and ambitious engineer from India, referred to at the 2019 UN Climate Action Summit when he said: 'Entrepreneurship and technological innovation – these are going to be the two major pillars ... if we are to fight climate change.'[66] He was followed on the panel by Greta Thunberg, who since then has risen to world fame, whereas nobody remembers Anurag Saha Roy.

Is nuclear power a technofix? Absolutely. Naomi Klein was right when she called it one of the 'Big Tech fixes to climate change'.[67] In fact, nuclear power turned out to be the *ultimate* technofix. Wars were avoided with the mere threat of an atomic bomb. Nuclear plants gave us better lives and a better environment all in one. The atom was banging on the door of political and economic powers, ready to bring more freedom and equality.

Whoa, wait! It wasn't supposed to go like this!

For those who had something to lose – influence, prestige, a worldview – nuclear power was so disruptive it angered them. Intellectuals questioned whether humanity could cope with a force with which we could do so much harm. The question they ignored was: *can humanity cope with a force that can do so much good?*

Facts mattered less and less, emotions didn't let themselves be put aside. A deep-rooted prejudice was looking for justification. Nuclear power would be too dangerous, would produce unmanageable waste, would be too expensive, would come too late…

They kept on saying it, again and again, with a pedantry they would never tolerate from their critics, combined with a hint of moral superiority. Nuclear power was not the answer, oh no – they knew better. They issued an impassioned, if rather non-specific, call for everything and everyone to change. They said it yesterday, they're saying it today, and they will say it again tomorrow.

The unprecedented potential of nuclear power is perhaps better understood by its opponents than its proponents. This potential became its weakness. A technological solution? *Ha, that would be really easy, wouldn't it?* No, society must be turned upside down, and the people – not them, of course, but *other* people – must reflect and repent.

It won't be easy. Greta Thunberg: 'In such an emergency as we're in right now, everyone needs to take more responsibility.'[68] Naomi Klein, at the press conference at the Vatican: 'To change everything, we need everyone.'[69] You see, that's the holy way.

Nuclear power was recognised early on as a technofix. Indeed, the word 'technofix' was coined in 1966 in recognition of the great value of nuclear power. Its originator was nuclear physicist Alvin Weinberg.[70] He meant it as a recommendation.

Lastly, three brief historical observations, as a prelude to some final reflections.

1. *Radioactivity was discovered at a time when many doctors didn't even wash their hands.*
2. *The atomic bomb first exploded in a world in which Polish soldiers had just been warding off German invasion on horseback, armed with lances and bayonets.*
3. *The first nuclear reactor was built when nobody owned a colour TV.*

It's obvious that nuclear power was always miles ahead of society at large. Perhaps we were never ready for something so revolutionary. Perhaps it's why we still struggle to understand it. Nuclear power didn't come 'too late' at all, as some say: it came too early.

Frederick Soddy was certainly thinking along those lines. The chemist who, along with Ernest Rutherford, demonstrated the mysterious transmutation of atoms in 1900, 'fervently hoped', according to his biographer, 'that the control of atomic energy would be postponed until society had become sufficiently mature to take responsibility for this achievement of scientific investigation, yet not postponed indefinitely, for then there would be an inevitable energy crisis.'[71]

When is society mature enough? Certainly not after the atomic bomb, Albert Einstein believed. 'The unleashed power of the atom has changed everything save our modes of thinking,' he wrote in 1946, 'and we thus drift toward unparalleled catastrophe.'[72] An unprecedented era of peace and prosperity had just begun, but even the most brilliant scientist was convinced we were falling headlong an era of destruction and decay.

And now? More than three-quarters of a century later, would Einstein see that we've learned to deal with the power of the atom? Would he even consider that we might be heading towards 'unparalleled catastrophe' if we *don't* harness that power?

With a revival of interest in nuclear power, even proponents recoil. Discovering the powerful position of opposing forces, some are getting cold feet and calling for innovation. In doing so, they condemn a long-proven technology to an eternal experiment.

Imagine that nuclear power did not exist. No Rutherford and Soddy. No Curie, Szilárd or Oppenheimer. No atomic bomb, no Hiroshima. No nuclear plants, no Chernobyl. What if someone today figured out how to safely produce an enormous amount of energy with few resources, which are available everywhere, without air pollution or CO_2 emissions and with only a little bit of recyclable waste, well separated from nature...

Surely we wouldn't hesitate to work on developing *that*, would we?

Epilogue

A feeling of awe

As I was working on this book, I stumbled upon an article about the 'worrying return' of nuclear power. At a climate conference, the industry had dared to present its product as 'a "safe and clean" way to generate electricity'. The journalist seemed bewildered that nuclear power was proposed as '*the* answer' to climate change.[1]

It was me who wrote those words. The year was 2000.

At that time, I wanted nothing to do with anything nuclear. Writing in *Ode*, an alternative magazine in the Netherlands I would ultimately serve as editor, I pointed to 'the overwhelming evidence' that nuclear power has 'nothing good to bring to people and nature'. Here's the last sentence of my piece: 'Now is the time to bring the nuclear industry down, before it can ruin the twenty-first century.'

Over the years, while covering environmental issues, I noticed how some of the most prominent voices within the movement, such as Stewart Brand, James Lovelock and George Monbiot, stood up to defend nuclear energy. Slowly, I came to accept it as a necessary evil. Nuclear power was a solution, indeed – far from ideal, but one which we couldn't exclude if we were serious about fighting global warming. Yet, I wasn't entirely convinced. Things could still go terribly wrong.

Right?

As recently as 2018, I wrote warily about nuclear power in my book on the transition from fossil fuels to renewables. On its final pages, I was forced to conclude that solar and wind do not contribute enough, for now, to drastically reduce carbon emissions. What then? I cautiously wondered if nuclear power was an option. 'If we fail to use the carbon-free energy from nuclear plants, we're missing a huge opportunity.'[2]

I was by no means effusive. 'Safety remains a thorny issue,' I wrote. The impact of an accident could be 'enormous', the problem with storing the waste was 'complex'. I added there was 'no public support' anyway, brushing the topic aside.

As a speaker at symposia on climate policy, I would sometimes suggest we should consider nuclear energy. In response, a few people would walk angrily out of the conference room. Once, someone shouted incessantly: 'You are a liar!' Interestingly, others applauded my suggestion. Such emotional responses never occurred when batteries or heat pumps were raised.

I needed to learn more, especially when politicians in the Netherlands started talking about including nuclear energy as part of climate policy. For the first time, I took a serious look at nuclear power. I read books, listened to podcasts and talked to experts. After a while it dawned on me: just about everything we think we know about nuclear power turns out to be wrong. Indeed, it's often exactly the opposite.

And then, Vladimir Putin's army invaded Ukraine. That day, 24 February 2022, Russian soldiers stormed the long-closed nuclear plant near Chernobyl. Soon after, weapons clashed at the nuclear plant in Zaporizhzhia. The Russians fired rocket launchers, the Ukrainians fired anti-tank missiles.

Immediately, warnings were issued that public health across Europe was at stake. Reports of increased radiation levels around Chernobyl became news. Russian armoured tanks had ploughed through the ground, releasing radioactive dust. We read about Russian soldiers digging trenches without any radiation protection, ignorant of the dangers because the army command had not warned them. They were reportedly taken to health clinics with symptoms of acute radiation sickness. It seemed that one death had already occurred.[3]

Soon it became clear that even the elevated radiation levels at Chernobyl were still very low and possibly manipulated.[4]

Articles about sick soldiers and one dead person were fake news. Thankfully, at Zaporizhzhia, the damage appeared to be limited.

Both sides accused the other of misconduct. Volodymyr Zelensky, the Ukrainian president, spoke of 'nuclear terror'.[5] Vladimir Putin stressed in a statement that 'the systematic shelling by the Ukrainian military of the territory of the Zaporizhzhia nuclear power plant creates the danger of a large-scale catastrophe'.[6]

Amid the war propaganda, Rafael Mariano Grossi, director general of the International Atomic Energy Agency (IAEA), spoke of 'the very real risk of a nuclear disaster that could threaten public health and the environment in Ukraine and beyond'.[7]

All the while, we had been alerted to the nuclear arsenal at the Kremlin's disposal. Is nuclear war imminent? Would Putin...? Nuclear plants and nuclear weapons came together. Who wouldn't get a little scared? I know I did.

While nuclear reactors are built to withstand bombs, missiles and shells, power grids can go down much more easily. When that happens, the reactor's cooling can stop. If the diesel generators also fail, and the batteries are dead, and the water storage tanks are broken, and the fire brigade cannot get the emergency cooling system to work, then, over time, a meltdown may occur, as happened at Fukushima in 2011.

While writing this book, I wondered: what would happen if a meltdown occurred at the nuclear plant near Zaporizhzhia?

You have read the book. You know that in the event of a meltdown, radiation may be released, after which for some local residents, someday, later in life, the theoretical risk of dying of cancer increases slightly. But what is the situation at Zaporizhzhia? Experts pointed out that the plant has a modern design. Each reactor is enclosed in a pressurised steel vessel, housed inside 'a massive reinforced-concrete containment structure'.[8] According to a 2016 report by the Joint Research Centre, the European Commission's science service, the reactors also have a system to keep the molten fuel on board in case of a meltdown.[9]

This means that Zaporizhzhia has two safety measures not in place at Fukushima. It means that it remains to be seen whether local residents in Zaporizhzhia will be exposed to radiation. If radioactivity is released at all, it is likely to be even less than in Fukushima. We can assume nobody will ever detect any ill health effects.

But that's not the whole story. We also know something else will happen – *all hell will break loose*. Journalists will think they are witnessing world history. Politicians will overprotect their citizens. Well-known opponents of nuclear power will come up with the worst scenarios. Activists will speak of countless deaths. Experts in the nuclear industry who built their careers on fear of radiation will make pompous claims about the unique dangers. Meanwhile, in Ukraine, men, women and children are dying in a war. *That* is the real drama.

Back in 2011, Japanese politicians were unprepared for a devastating tsunami, unable to protect their citizens from the natural disaster. They quickly shifted attention to saving them from the consequences of a damaged nuclear plant. A full decade later, Europe's politicians proved unprepared for Putin's aggression. The damaged nuclear plant gave them an opportunity to show decisive action. Needless to say, the turmoil was not inconvenient for power companies selling electricity produced by natural gas, coal, biomass, hydro, wind or solar.

Just as the original, Dutch edition of my book went to print in September 2022, the IAEA released a report on the latest inspections at Zaporizhzhia, which by now no longer supplied the grid.[10] In the spotlight of international attention, top executive Grossi seized the opportunity to shine on the world stage. Addressing the UN Security Council, he said, 'Something very, very catastrophic could take place.'[11]

One newspaper proclaimed that a 'nuclear catastrophe looms',[12] another that a 'nuclear disaster looms'.[13] That's the kind of news that sticks. The report itself said that all safety systems were operational. Maybe the truth about nuclear power is so shockingly unsensational that it's rarely noticed.

As I shifted further towards the proponent side, something else happened: I gained a better understanding of the suspicion so many people have about nuclear. I became fascinated by the way thinking about nuclear energy has been influenced by its history and by social trends, financial interests and political choices.

That's why I didn't want to write a predictable book outlining a case for nuclear power. I wanted to offer the incredible story of a technology that was misunderstood from the beginning. A story of life and death, of hope and fear. Rather than providing an answer to the question of what we should do with nuclear energy, another question seemed at least as appropriate: *what does nuclear power do to us?* While writing this book, that question would remain in the back of my mind.

When I told my wife about this new angle, she asked: 'And what does nuclear power do to you?'

I couldn't answer.

Now that the book is finished, I have a better idea of the answer. I find nuclear power a great miracle. I'm captivated by the wonders of such technical ingenuity, even though, quite frankly, I'm only moderately interested in the technology itself. Whenever we drive past a nuclear plant, for a holiday on the Zeeland coast or beyond the border near Antwerp, I get a feeling of awe – the same kind of feeling that creeps up on me when the home stadium of my football club Feyenoord comes into view. I also feel it when visiting a cathedral.

Yes, I'm guilty of being struck by the enthusiasm of the 1950s. But there's one difference. Back then, nuclear power was an uncertain adventure; today, we lean on more than 70 years of experience.

Still, I too am left with questions. Is a nuclear plant so special that it justifies all kinds of unique regulations, or is it nothing more than another way to produce energy? Is a country's energy system so important that everyone should have a say in it, or is it so important that it's better left to experts? Is energy something that should be left to the private market, or is it, as in the old days, a public utility to be provided by the government?

But wait, what if politics is dominated by people who do not value science or technical expertise? And what if no effort is made to firmly establish in society a broad support for nuclear power? A new government may renounce it just like that, whereas a choice for nuclear power should be one for the long, yes, the very long term. Perhaps the choice is only comparable to the decision made by the Dutch, hundreds of years ago, to build and maintain dykes to keep the water out.

Can we really muster the will to build, keep and nurture nuclear plants?

Finally, is now not the time to end the fear of nuclear power?

Acknowledgements

Many people thought and read along with me while I was working on this book. It's thanks to them that I've learned new insights and avoided some embarrassing mistakes. And thanks to them, writing the book was much more fun.

First of all, there was Jaffe Vink, who taught me how to write. Jaffe, thank you for helping me write the book I wanted to write.

Here's a big thank you to the two most faithful readers: Joris van Dorp and Lars Roobol. Both went out of their way to educate me. Others whom I would like to thank for reading along and commenting on the manuscript were Roger Blomqvist, Maarten Boudry, Remko van Broekhoven, Manuel Sintubin and George Verberg. When it came to nuclear weapons, Bertjan Verbeek read for accuracy. If any errors are still in the text, they are all mine.

I'm also grateful for all kinds of experts who provided me with ideas and perspectives: Take Aanstoot, Sweden; Adam Błażowski, Poland; Ralf Bodelier, Wouter van Caspel, Yonis Le Grand, Jules Lavalaye, Olguita Oudendijk, Behnam Taebi, Mirjam Vossen and Gijs Zwartsenberg, the Netherlands; Robert Bryce, Joshua Goldstein, Madison 'Madi' Hilly, Todd Moss, Mark Nelson, Paris Ortiz-Wines and Richard Rhodes, United States; Andreas Fellner, Austria; Simon Friederich, Stephen Milder and Amardeo Sarma, Germany; Luís Guimarãis, Portugal; Tyrone D'Lisle, Australia; Mark Lynas and Geraldine Thomas, United Kingdom; Princy Mthombeni, South Africa; Rauli Partanen and Tea Törmänen, Finland; and Rob De Schutter, Belgium.

Special thanks go to Michael Shellenberger, who confronts the beliefs of nuclear opponents just as fearlessly as he does those of proponents. His original thinking is an enduring inspiration.

I've very much enjoyed working with Lewis Ward and Catherine Best, both meticulous copy-editors, and with the folks at Bloomsbury Publishing, most notably Sarah Lambert.

Finally, the most important of all. Neeltje, thank you for your support and your love. Puk, Otje and Kobus: *Daddy has finished!*

A Note on Sources

For general information on anything from trends in energy use or carbon emissions to the health effects of radiation, I've turned to reports from established institutions, such as the International Energy Agency (IEA), the International Atomic Energy Agency (IAEA), the Intergovernmental Panel on Climate Change (IPCC), the United Nations Scientific Committee on the Effects of Atomic Radiation (UNSCEAR), the World Health Organization (WHO) and the like. I've also used online platforms with outstanding data gathering and visualisation, such as Statista and Our World in Data. The World Nuclear Association, which promotes nuclear power and supports companies in the industry, offers much relevant information, for example on the most current plans for building nuclear reactors or legislation on waste management.

Soon after I decided to write a book about nuclear energy, the COVID-19 pandemic broke out. With no social life to speak of, I immersed myself in books – lots of books. Perhaps the most influential title on my thinking about nuclear energy is *The Rise of Nuclear Fear* by Spencer Weart (Harvard University Press, 2012), a revised version of his 1988 book *Nuclear Fear*. This cultural history of nuclear energy taught me how today's emotions about the technology, both positive and negative, go back to the atomic bomb – even back to the time when nuclear energy didn't yet exist.

This insight inspired me to look into nuclear weapons much more closely than many proponents of nuclear energy are willing to do. The standard work is *The Making of the Atomic Bomb* by Richard Rhodes (Simon & Schuster, 1986). At nearly 1,000 pages, this is by far the longest page-turner I have ever read. If you're more into graphic novels, check out

The Bomb: The Weapon That Changed the World by Alcante and Laurent-Frédéric Bollée, with wonderful illustrations by Denis Rodier (Abrams ComicArts, 2023).

Other relevant books about the scientific discoveries and events leading up to the atomic bomb are *American Prometheus: The Triumph and Tragedy of J. Robert Oppenheimer* by Kai Bird and Martin Sherwin (Vintage Books, 2005) and *The Age of Radiance: The Epic Rise and Dramatic Fall of the Atomic Era* by Craig Nelson (Scribner, 2014). The Atomic Heritage Foundation, dedicated to the preservation and interpretation of the Manhattan Project, has a fantastic website (ahf .nuclearmuseum.org), including oral histories and virtual tours of the sites.

On nuclear weapons in general, I recommend *Command and Control: Nuclear Weapons, the Damascus Accident, and the Illusion of Safety* by Eric Schlosser (Penguin Books, 2014). The idea of deterrence was pretty new to me. I learned more about this in *Deterrence Before Hiroshima* by George Quester (Taylor & Francis, 1986). Most of the inspiration for my chapter on nuclear weapons comes from *Atomic Obsession: Nuclear Alarmism from Hiroshima to Al-Qaeda* by John Mueller (Oxford University Press, 2010), a fantastic book that challenges conventional wisdom about nuclear weapons.

Some noteworthy books on nuclear energy in general are *Power to Save the World: The Truth About Nuclear Energy* by Gwyneth Cravens (Vintage Books, 2007); *Atomic Awakening: A New Look At The History And Future Of Nuclear Power* by James Mahaffey (Pegasus, 2009); *Fallout: Disasters, Lies, and the Legacy of the Nuclear Age* by Fred Pearce (Beacon Press, 2018); and *Atoms and Ashes: From Bikini Atoll to Fukushima* by Serhii Plokhy (Penguin, 2022).

Books that discuss nuclear energy specifically in today's context of climate change are *Nuclear 2.0: Why a Green Future Needs Nuclear Power* by Mark Lynas (UIT Cambridge, 2014), and two self-published books by Rauli Partanen and Janne Korhonen: *Climate Gamble: Is Anti-Nuclear Activism Endangering*

Our Future (2015) and *The Dark Horse: Nuclear Power and Climate Change* (2020). A history of our knowledge about climate change can be found in *The Discovery of Global Warming*, another great book by Spencer Weart (Harvard University Press, 2008).

To gain a better understanding of energy in general, I recommend *A Question of Power: Electricity and the Wealth of Nations* by Robert Bryce (Public Affairs, 2020), *Energy: A Human History* by Richard Rhodes (Simon & Schuster, 2018) and *The Age of Energy: Understanding Growth, Prosperity, and Environmental Destruction* by Rauli Partanen and Aki Suokko (self-published, 2022). If you enjoy listening to podcasts, I recommend Decouple, Power Hungry and Titans of Nuclear. I've listened to numerous episodes, learning about perspectives from a wide variety of experts.

To inform myself about the details of what happened during nuclear accidents of the past, I learned much from *Atomic Accidents: A History of Nuclear Meltdowns and Disaster: From the Ozark Mountains to Fukushima* by James Mahaffey (Pegasus, 2015). On the events in Chernobyl, one book stands out: *Midnight in Chernobyl: The Untold Story of the World's Greatest Nuclear Disaster* by Adam Higginbotham (Transworld, 2019). On Fukushima, a sober account is offered in *Meltdown: Inside the Fukushima Nuclear Crisis* by Yoichi Funabashi (Brookings Institution Press, 2021).

Finally, radiation is explained well in *Radiation: What It Is, What You Need to Know* by Robert Peter Gale and Eric Lax (Vintage Books, 2013) and *Strange Glow: The Story of Radiation* by Timothy Jorgensen (Princeton University Press, 2016).

I feel much gratitude for the excellent work of so many fine authors.

Notes

Prologue

1 M. Visscher, '"Het verzet tegen kernenergie is echt krankzinnig"', *de Volkskrant*, 18 November 2017, tinyurl.com/yc2ueyfk.

2 The exact amount of German government spending on the energy transition has been assessed as unclear. According to experts, the total will amount to more than €520 billion (£444 billion) by 2025. See, among others: H. Kuittinen and D. Velte (2018), *Mission-oriented R&I Policies: Case Study Report Energiewende*, European Union, tinyurl.com/7thhb3rb.

3 Statista: 'Power sector emissions in Germany from 2000 to 2021', tinyurl.com/27h4c9n5

4 Among Europe's top seven most polluting coal plants, five were found to be in Germany. See: K. Gutmann *et al.* (2014), *Europe's Dirty 30: How the EU's Coal-fired Power Plants Are Undermining Its Climate Efforts*, CAN Europe, tinyurl.com/2jx3ezhw.

5 S. Amelang *et al.*, 'Germany's energy use and emissions likely to rise yet again in 2017', *Clean Energy Wire*, 13 November 2017, tinyurl.com/23k7xktv.

6 World Resources Institute: 'CO_2 emissions – European Union, Germany', tinyurl.com/yc2yk4b8.

7 Our World in Data: 'Energy Production and Consumption', tinyurl.com/y2c8shyn.

8 Our World in Data: 'Energy Production and Consumption', tinyurl.com/y2c8shyn.

9 J. Hansen, 'The new testimony before Congress', *Grist*, 24 June 2008, tinyurl.com/bp8rsdkc.

10 'Direct Testimony of James E. Hansen, State of Iowa Before the Iowa Utilities Board', 5 November 2007, tinyurl.com/2yw33dcx.

11 Press conference on YouTube: tinyurl.com/4sunxjx5.

12 Statista: 'Median construction time required for nuclear reactors worldwide from 1981 to 2020', tinyurl.com/4f479pe2.

13 The World Nuclear Association offers the most recent figures on global nuclear energy production: 'Nuclear Power in the World Today', tinyurl.com/yvveyfh9.

14 Eurostat: 'What is the source of the electricity we consume?', tinyurl.com/yjxc8npy.

15 The World Nuclear Association offers the most recent figures on construction plans of nuclear reactors around the world: 'Plans For New Reactors Worldwide', tinyurl.com/mpa3tpuu.

16 The World Nuclear Association: 'Reactor Database', tinyurl.com/yc327xmu.

17 Our World in Data: 'Share of electricity production from nuclear', tinyurl.com/4p4dtvx3.

18 H. de Coninck *et al.* (2018), *Global Warming of 1.5 °C: An IPCC Special Report on the impacts of global warming of 1.5 °C above preindustrial levels and related global greenhouse gas emission pathways, in the context of strengthening the global response to the threat of climate change, sustainable development, and efforts to eradicate poverty*, Cambridge University Press, pp. 313–444, tinyurl.com/muewf6nk.

Chapter 1

1 Video showing J. Robert Oppenheimer talking about the Trinity test of the atomic bomb: tinyurl.com/4byjnnpw.

2 That question was asked by George Caron, the tail gunner. Source: W. Lawrence (1946), *Dawn Over Zero: The Story of the Atomic Bomb*, Knopf, p. 220. Quoted in: R. Rhodes (1986), *The Making of the Atomic Bomb*, Simon & Schuster, p. 707.

3 Robert Lewis talked about this in a documentary by Steven Okazaki: *White Light/Black Rain: The Destruction of Hiroshima and Nagasaki* (2007). The clip can be seen at: tinyurl.com/2suc668a.

4 This analogy comes from: R.P. Gale and E. Lax (2013), *Radiation: What It Is, What You Need to Know*, Vintage Books, p. 12.

5 M. Howorth (1958), *Pioneer Research on the Atom: Rutherford and Soddy in a Glorious Chapter of Science: The Life Story of Frederick Soddy*, New World, p. 84. Quoted in: S. Weart (2012), *The Rise of Nuclear Fear*, Harvard University Press, p. 3.

6 Correspondence with William Cecil Dampier, 26 July 1903. Cited in: S. Weart (1988), *Nuclear Fear: A History of Images*, Harvard University Press, p. 18.

7 M. Howorth (1953), *Atomic Transmutation: The Greatest Discovery Ever Made*, New World, p. 95. Quoted in: Rhodes (1986), p. 44.

8 This colleague was Patrick Blackett. The year was 1925.

9 Inaugural address, meeting of the Society of Friends of Radio, 1 March 1926, tinyurl.com/ywdxzba2.

10 H. Kragh (2002), *Quantum Generations: A History of Physics in the Twentieth Century*, Princeton University Press, p. 770.

11 Quoted in: J.L. Heilbron (2003), *Ernest Rutherford: And the Explosion of Atoms*, Oxford University Press, p. 118.

12 A.S. Eve (1939), *Rutherford: Being the Life and Letters of the Rt. Hon. Lord Rutherford, O.M.*, Macmillan, p. 374.

13 G. Jenkin, 'Atomic Energy is "Moonshine": What did Rutherford Really Mean?', *Physics in Perspective*, vol. 13, 2011, pp. 128–45, tinyurl.com/2p8j2m23.

14 W. Kaempffert, 'Rutherford Cools Atom Energy Hope', *New York Times*, 12 September 1933, p. 1, tinyurl.com/2e39kypj.

15 Fritz Strassmann was in the laboratory with Hahn at the time; Meitner was in Denmark and had the insight after discussing a letter from Hahn with her cousin, Otto Frisch. The Nobel Prize for this work went to Hahn.

16 W. Lanouette and B. Silard (1992), *Genius in the Shadows: A Biography of Leo Szilárd: The Man Behind The Bomb*, Charles Scribner's Sons, p. 199.

17 Declaration by Roosevelt, 1 September 1939, tinyurl.com/ u2v7rk6a.
18 'Enrico Fermi Dead at 53; Architect of Atomic Bomb', *New York Times*, 29 November 1954, p. 1, tinyurl.com /5b4r5sam.
19 J. Mahaffey (2009), *Atomic Awakening: A New Look at the History and Future of Nuclear Power*, Pegasus, p. 108.
20 Interview, CBS Reports, 2 April 1960, tinyurl.com/5a95u4nj.
21 Statement, 14 August 1945, tinyurl.com/mufhh9az.

Chapter 2

1 'Prins Bernhard opent Het Atoom', *Het Vrije Volk*, 29 June 1957, p. 16, tinyurl.com/ykjajta3.
2 Polygoon newsreel, shown as part of the documentary 'Geloof in kernenergie' by television programme *Andere Tijden*, 17 May 2005, tinyurl.com/2e6fupnd.
3 Government programme, 2 September 1956, included as an appendix in: J.W. Brouwer and P. van der Heiden (eds.) (2004), *Het kabinet-Drees IV en het kabinet-Beel II, 1956–1959: Het einde van de rooms-rode coalitie*, Sdu Uitgevers, p. 322, tinyurl.com/5wbmtma3.
4 'Nota inzake de Kernenergie (Opwekking van electriciteit door middel van kernenergie)', sent to parliament on 3 July 1957, tinyurl.com/ydbrh67r.
5 Eisenhower's speech, on 8 December 1953, including erasures: tinyurl.com/2p8dzdny. A video recording can be seen here: tinyurl.com/yhns7vws.
6 L. Velleman, 'Kernenergie kan voorkomen dat in het jaar 2000 millioenen omkomen', *Het Vrije Volk*, 26 June 1954, p. 3, tinyurl.com/3pu8dfz3.
7 'Dr. Plesman: K.L.M. heeft geen reden tot klagen', *Java-Bode*, 20 November 1953, p. 1, tinyurl.com/2cux6dnr.
8 The family name is Van der Vorm. J. Dohmen, 'De geheime bron van de miljoenen die de familie Van der Vorm uitdeelt', *NRC Handelsblad*, 8 September 2023, tinyurl.com/2xxw2vhc.

9 S. Slagter (1959), *De mens in maatschappij, techniek en cultuur*, Uitgeversmaatschappij W. de Haan, p. 87.

10 'Our Friend the Atom', tinyurl.com/2p8trbv6.

11 Quoted in: W. Cornelisse (1995), *Waar licht is, is vreugde: Een eeuw gemeentelijke energievoorziening in Amsterdam*, N.V. Energiebedrijf Amsterdam, p. 105. Sources cited are: R. Houwink, 'Onze energiebehoeften in de toekomst', *Mens en Onderneming*, November 1957, and editions of *De Koppeling*, December 1957 and March 1959.

12 'Aarde wordt steeds meer een broeikas', *de Volkskrant*, 29 December 1956, p. 7, tinyurl.com/2p9anwm8.

13 J. Monnet (2015), *Memoirs*, Third Millennium Publishing, p. 418, tinyurl.com/ywp9ss66.

14 'Tilburg heeft primeur van W.-Europa voor atoomkracht in vredesdienst', *Het Nieuwsblad van het Zuiden*, 31 March 1955, p. 7, tinyurl.com/2p8p4wkh.

15 'Eén dag de wereld rond', *Het Nieuwsblad van het Zuiden*, 2 April 1955, p. 4, tinyurl.com/ehjv7mhj.

16 T. Zoellner (2010), *Uranium: War, Energy and the Rock That Shaped the World*, Penguin Books, p. viii.

17 The composer with a telescope was William Herschel.

18 The prospector was Robert Rich Sharp.

19 The geologist was Charles Steen.

20 Lewis Strauss, chairman of AEC, at a press conference, 31 March 1954, recorded in 'Hearing Before the Subcommittee on Security of the Joint Committee on Atomic Energy, Congress of the United States, Eighty-fourth Congress, First Session on AEC–FCDA Relationship', 24 March 1955, p. 8, tinyurl.com/2s46vf7n.

21 'Het atoom te kijk gesteld' and 'Vijf technici klommen naar niet ontplofte A-bom', *De Telegraaf*, 29 June 1957, p. 3, tinyurl.com/yerkj38r.

22 B. van der Boom (2000), *Atoomgevaar? Dan zeker B.B.: De geschiedenis van de Bescherming Bevolking*, Sdu Uitgevers, pp. 26, 55–6. Quoted in: D. van Lente, 'Temmen en

verontrusten: De atoombom in de stripverhalen Blake en Mortimer en Tom Poes (1946–1960)', *Gewina*, vol. 29, no. 4, 2006, p. 15, tinyurl.com/438b7fuk.

Chapter 3

1 G.W. Ndoye, 'Retour sur Malville, Chaim Nissim a coeur ouvert', *Continent Premier*, 4 July 2007, tinyurl.com/8nnpxhkv.

2 G. Schwab (1963), *Dance With the Devil: A Dramatic Encounter*, G. Bles. The chapter on nuclear power is available online: tinyurl.com/3dbrykwv.

3 'Reckless play with nature', *de Volkskrant*, 1 October 1960, p. 8, tinyurl.com/y8v3bjva.

4 'Nucleus': J. Radkau (2000), *Nature and Power. A World History of the Environment*, Beck, p. 304. 'A significant impetus': J. Radkau, 'Eine kurze Geschichte der deutschen Antiatomkraftbewegung', Bundeszentrale für politische Bildung (BPB), 10 November 2011, tinyurl.com/2bef2hdz.

5 '"Duivel" nu verknipt tot "lot"', *Trouw*, 8 January 1975, p. T5/K7, tinyurl.com/3dx4m7xt.

6 G. Schwab, 'Die Menschheit am Neubeginn', *Gesundheitsberater*, January 1992, p. 11. Cited in: J. Ditfurth (1996), *Entspannt in die Barbarei: Esoterik, (Öko-)Faschismus und Biozentrismus*, Konkret Literatur Verlag, p. 39, tinyurl .com/ws2rhd92.

7 The chair's name was Werner Georg Haverbeck; the organisation was Collegium Humanum. See: 'Schäuble verbietet rechtsextreme Organisationen', *Die Welt*, 7 May 2008, tinyurl.com/bddvkjrf.

8 His name was Max Otto Bruker. See: J. Ditfurth, 'Braunes Müsli: "Ernährungspapst" mit Neonazikontakten', *Max*, April 1994, p. 201.

9 Her name: Ursula Haverbeck-Wetzel. See: 'Germany's "Nazi Grandma" given jail term for Holocaust denial', BBC, 29 November 2017, tinyurl.com/y7jw9tc9.

10 E. Sternglass, 'The Death of All Children', *Esquire*, 1 September 1969, tinyurl.com/yc5tfhdm.

11 The analysis of shifting attention, based on Sigmund Freud's *Verschiebung* theory, comes from S. Weart (1988), *Nuclear Fear: A History of Images*, Harvard University Press, p. 212.

12 D. McTaggart and R. Hunter (1978), *Greenpeace III: Journey into the Bomb*, Collins, p. 106.

13 Sources: various Greenpeace websites. 'Dirty and dangerous' from the international website, tinyurl.com/4np6msv8; 'disastrous catastrophes' from the Dutch website, tinyurl .com/44u4evau; 'huge amounts of hazardous waste' from the UK website, tinyurl.com/h34uhtc2; and 'no place' from the international website, tinyurl.com/4avk3jku.

14 S. Rothwell, 'antinuclear movement', *Encyclopaedia Britannica* website, 8 December 2023, tinyurl.com/4fwsw2z2.

15 A. Weinberg (1994), *The First Nuclear Era: The Life and Times of a Technological Fixer*, Springer, p. 271.

16 C.P. Snow's so-called Rede Lecture on 7 May 1959 (tinyurl .com/2xwvpcay) was based on his article: 'The Two Cultures', *New Statesman*, 6 October 1956, tinyurl.com/ mrxwjxf3.

17 Lecture to the Society for Democratic Integration in Industry in London, October 1962, 'Modern Industry in the Light of the Gospel', tinyurl.com/2s3nehfu.

18 E.F. Schumacher (2011, originally 1973), *Small Is Beautiful: A Study of Economics as if People Mattered*, Vintage Books, p. 116, tinyurl.com/2p9ds6ue.

19 Lecture, October 1962.

20 Schumacher (1973), p. 102.

21 Schumacher in conversation with Satish Kumar, founder of the Schumacher College, in: E.F. Schumacher (1997), *This I Believe and Other Essays*, Green Books, p. 8, tinyurl.com /3k2fd4vj.

22 The essay 'Buddhist Economics' (tinyurl.com/yckz3evx) first appeared in G. Wint (ed.) (1966), *Asia: A Handbook*, Praeger, and is included in Schumacher (1973), p. 38.

23 From the 1977 obituary in *The Times*, written by Barbara Ward, quoted in: R. McCrum, 'E.F. Schumacher: Cameron's choice', *Guardian*, 27 March 2011, tinyurl.com/y6m9kcp9.

24 *Times Literary Supplement*, 6 October 1995, p. 39.

25 Schwab (1963), chapter 2, tinyurl.com/24nkarz6.

26 Schumacher (1973), p. 24.

27 P. Ehrlich, 'Machine guns and idiot children', *Not Man Apart*, vol. 5, no. 18, September 1975. The magazine was a publication of Friends of the Earth.

28 'The Plowboy Interview: Amory Lovins, Energy Analyst', *Mother Earth News*, 1 November 1977, tinyurl.com/mr2bzcys.

29 R. Adams, 'Smoking gun: Robert Anderson provided initial funds to form Friends of the Earth', *Atomic Insights*, 4 August 2013, tinyurl.com/mr95b57j.

30 Such as in the documentary 'Geloof in kernenergie' by television programme *Andere Tijden*, 17 May 2005, tinyurl.com/2e6fupnd.

31 Möller's story is recorded in her own words in: 'Onderzoeksverslag naar buitenparlementaire acties omtrent het kernenergie-vraagstuk in Nederland voor de Universiteit van Amsterdam', by Werkgroep 2 Politicologie, May 2008, tinyurl.com/43yekdv2.

32 'Geloof in kernenergie', 2005.

33 M. van Calmthout, 'Interview: Wim Turkenburg', *de Volkskrant*, 7 April 2012, tinyurl.com/37ta7bv6. Note: the newspaper said 'Copenhagen', not Stockholm. I made this correction after correspondence via email with Turkenburg, 6 January 2022.

34 'Kernenergienota', September 1972, compiled by Werkgroep Kernenergie, released by Vereniging Milieudefensie, tinyurl.com/tv6pkk7s.

35 Schumacher (1973), p. 110.

36 Schumacher (1973), p. 112.

37 Schumacher (1973), pp. 101–2.

38 Quoted in a review of *Nuclear Energy: Promises, Promises* by George Weil, in: W. Patterson, 'George Weil, from activator

to activist', *New Scientist*, 30 November 1972, p. 520, tinyurl
.com/2p877w6c.

39 Roger Ebert, a *Chicago Sun-Times* film critic who was the
first in his profession to be awarded a Pulitzer Prize for his
work, tinyurl.com/4huyejx2.

40 V. Canby, 'Film: Nuclear Plant Is Villain in "China
Syndrome": A Question of Ethics', *New York Times*, 16
March 1979, p. 0, tinyurl.com/567hnw3w.

41 Quotation from John Taylor, who headed Westinghouse's
nuclear energy department, in: D. Burnham, 'Nuclear
Experts Debate "The China Syndrome"', *New York Times*,
18 March 1979, section D, p. 1, tinyurl.com/3a2c69tx.

42 W. Laurence, 'Drama of the Atomic Bomb Found Climax
in July 16 Test', *New York Times*, 25 September 1945, tinyurl
.com/mryuf2a4.

43 Quote from General Bruce Holloway, former commander-
in-chief of Strategic Air Command, in: D. Ford, 'The
button', *New Yorker*, 1 April 1985, p. 49.

44 H. Gusterson (1996), *Nuclear Rites: A Weapons Laboratory at
the End of the Cold War*, University of California Press,
p. 163.

45 Documentary, *If You Love This Planet*, 1982, tinyurl.com/
mr3atz43.

Chapter 4

1 The entire text of the radio message, broadcast on 27 April
1986 at 13.10, is available in: A. Sidorchik, 'Deadly
Experiment: Timeline of the Chernobyl Disaster', *Argumenty
i fakty*, 26 April 2016, tinyurl.com/yn9v6zcb.

2 The story of the traffic controller, Maria Protsenko, can be
found in: A. Higginbotham (2019), *Midnight in Chernobyl:
The Untold Story of the World's Greatest Nuclear Disaster*, Simon
& Schuster, pp. 169–71.

3 L. Whittington, '"2,000 Die" in Nukemare; Soviets Appeal
for Help as N-Plant Burns out of Control', *New York Post*, 29
April 1986. Cited in: Higginbotham (2019), p. 181.

4 G. Hawtin, 'Report: 15,000 Buried in Nuke Disposal Site', *New York Post*, 2 May 1986. Cited in: Higginbotham (2019), p. 181.

5 *Igor, Child of Chernobyl*, 1995, tinyurl.com/5x2rvtaz.

6 R. Schulze, 'Und kein bisschen müde', *Frankfurter Allgemeiner Zeitung*, 23 March 2008, tinyurl.com/mrx8h2we.

7 G. Pausewang, 'Solange ich lebe, werde ich warnen', *Der Spiegel*, 17 March 2011, tinyurl.com/2663eb8d.

8 IMDB: tinyurl.com/2p93n2uv.

9 'Swedish Chief Assails Nuclear Power', *New York Times*, 18 August 1986, p. 6, tinyurl.com/2s4cd2xu.

10 M. Gorbachev, 'Turning Point at Chernobyl', *Project Syndicate*, 14 April 2006, tinyurl.com/5c2seh5a.

11 An oft-cited quote by energy expert Mikhail Styrikovich from a 1980 article in *Ogonyok* magazine.

12 The *Soviet Life* article was picked up after the Chernobyl accident, including in: 'Odds of Meltdown "One in 10,000 Years," Soviet Official Says', *The Associated Press*, 29 April 1986, tinyurl.com/93epp229.

13 Higginbotham (2019), p. 87.

14 Speech, 14 May 1986, tinyurl.com/yckkjdk8.

15 Nikolay Fomin, chief engineer and deputy director, was also sent to the penal camp.

16 Quote is by Nikolaj Steinberg, in: P.P. Read (1993), *Ablaze: The Story of the Heroes and Victims of Chernobyl*, Random House, p. 324. Cited in: Higginbotham (2019), p. 340.

17 Higginbotham (2019), p. 231.

18 The following paragraphs are based on The Chernobyl Forum (2005), *Chernobyl's Legacy: Health, Environmental and Socio-Economic Impacts*, tinyurl.com/36ydhdm5.

19 The Chernobyl Forum (2005), p. 16.

20 D. Schwartz, 'Craig Mazin's Years-Long Obsession with Making "Chernobyl" Terrifyingly Accurate', *Vice*, 3 June 2019, tinyurl.com/2wu8kd4w.

21 See, e.g.: LaCapria, K., 'The Chernobyl "Bridge of Death"', Truth or Fiction?, 6 June 2019, tinyurl.com/333nmr5j.

22 Footage of the collision on YouTube: tinyurl.com/2p8wssk6.

23 D. Stover, 'The human drama of Chernobyl', *Bulletin of the Atomic Scientists*, 5 May 2019, tinyurl.com/yc4paejy.

24 'Atom plant gets all clear: pile "under control"', *Shields Daily News*, 12 October 1957, p. 12, tinyurl.com/52btwxnk.

25 From the back cover of K. Brown (2019), *Manual for Survival: A Chernobyl Guide to the Future*, W.W. Norton & Co.

26 Brown (2019), p. 311.

27 M. Shellenberger, 'Interview of Kate Brown, author of *Manual for Survival*', 12 March 2019, tinyurl.com/3xwhu4e7.

28 Debate between Helen Caldicott and George Monbiot, *Democracy Now!*, 30 March 2011, tinyurl.com/3wuk9e62.

29 E. Aronova, 'Nuclear Fallout', *Science*, vol. 363, no. 6,431, p. 1,044, 8 March 2019, tinyurl.com/2h3utxdx.

30 S. Schmid, 'Chernobyl: data wars and disaster politics', *Nature*, vol. 566, pp. 450–1, 28 February 2019, tinyurl.com /3ds7ma9j.

31 'A view from the bridge: The tragedy of Chernobyl', *The Economist*, 9 March 2019, tinyurl.com/2p926udc.

Chapter 5

1 The UK Health Security Agency offers solid information on radiation. The doubling of radiation exposure in these examples is based on a map showing levels of radon in the UK, and can be found at tinyurl.com/5364uw52.

2 The UK Health Security Agency puts the fallout radiation from both nuclear weapons and nuclear accidents at 0.2 per cent of the annual average of 2.7 mSv. See: tinyurl.com /4pd5unfj.

3 The average radiation level resulting from medical diagnostics is distorted, as most people rarely deal with X-rays and CT scans while a small proportion deal with them a lot. The UK Health Security Agency puts the average radiation dose from medical diagnostics in the UK at 0.4 mSv per year. See: tinyurl.com/4pd5unfj.

4   A. Einstein, 'Über das Relativitätsprinzip und die aus demselben gezogenen Folgerungen', *Jahrbuch der Radioaktivität und Elektronik*, 4, 1907, pp. 411–62, tinyurl.com/nh5ezr8d. The equation E=mc² builds on a 1905 study: A. Einstein, 'Ist die Trägheit eines Körpers von seinem Energieinhalt abhängig?', *Annalen der Physik*, 18, pp. 639–41, tinyurl.com /48dbt45s.

5   T. Mann (1999, original 1927), *The Magic Mountain*, Random House, p. 214.

6   Mann (1927), p. 216.

7   The entrepreneur behind Radithor was William Bailey; the deadly victim was Eben Byers.

8   The story of the women has often been described, such as in K. Moore (2018), *The Radium Girls*, Simon & Schuster.

9   'Edison Fears Hidden Perils of the X-Rays', *New York World*, 3 August 1903, p. 1, tinyurl.com/mu838v3y.

10 R. Rhodes (1986), *The Making of the Atomic Bomb*, Simon & Schuster, p. 214.

11 Paracelsus, 'Die dritte Defension wegen des Schreibens der neuen Rezepte', *Septem Defensiones*, 1538, tinyurl.com /2kwfv3f5.

12 'Hearing Before the Special Committee on Atomic Energy, Seventy-ninth Congress, 27 November–3 December 1945, part 1', p. 37, tinyurl.com/3n2d5dcs.

13 Website, UK Health Security Agency, tinyurl.com/ yztdr6w4.

14 See, e.g.: T. Tollefsen *et al.* (2017), *European Atlas of Natural Radiation*, European Commission/Joint Research Centre, tinyurl.com/2f5tdmxp.

15 M. Ghiassi-nejad *et al.*, 'Very high background radiation areas of Ramsar, Iran: preliminary biological studies', *Health Physics*, vol. 82, no 1, pp. 87–93, January 2002, tinyurl.com /mpn8nrrv.

16 UNSCEAR, 'Sources, Effects and Risks of Ionizing Radiation, UNSCEAR 2013 Report, vol. 1: Report to the

General Assembly, Scientific Annex A', p. 8, 2014, tinyurl
.com/mppju5c9.

17 UNSCEAR, 'Sources and Effects of Ionizing Radiation,
    UNSCEAR 2008 Report, vol. 2: Effects, Scientific Annexes
    C, D and E', 2011, tinyurl.com/bdedezxn.

18 UNSCEAR (2011), p. 54.

19 Website, UK Health Security Agency, tinyurl.com/4pd5unfj.

20 Quote is by Zbigniew Jaworowski, former chairman of
    UNSCEAR, cited in: D. Bodansky (1996), *Nuclear Energy:
    Principles, Practices, and Prospects*, American Institute of
    Physics, p. 100, tinyurl.com/2p9bbm2y.

Chapter 6

1   N. Fukue, 'Japan nuclear refugees face dilemma over returning
    home', *AFP News*, 21 July 2015, tinyurl.com/23bu7kaz.

2   Statement, 11 March 2011, tinyurl.com/t3x48xjb.

3   Y. Funabashi (2021), *Meltdown: Inside the Fukushima Nuclear
    Crisis*, Brookings Institution Press, p. 35.

4   The close associate is Kenichi Shimomura. See: Funabashi
    (2021), p. 35.

5   The Deputy Chief Cabinet Secretary was Tetsuro Fukuyama.
    See: Funabashi (2021), p. 82.

6   R. Harding, 'Fukushima nuclear disaster: Did the evacuation
    raise the death toll?', *Financial Times*, 11 March 2018, tinyurl
    .com/2y9mttcf.

7   Fukue (2015).

8   UNSCEAR, 2011.

9   The Chernobyl Forum (2005), p. 36.

10 Documentary, *Living with Chernobyl: The Future of Nuclear
    Power*, 2007, tinyurl.com/2dwv79pc.

11 UNSCEAR, 'Sources, Effects and Risks of Ionizing
    Radiation, UNSCEAR 2013 Report, vol. 1: Report to the
    General Assembly', 2014, tinyurl.com/mppju5c9; and
    UNSCEAR, 'Sources, Effects and Risks of Ionizing
    Radiation, UNSCEAR 2020/2021 Report, vol. 2: Scientific
    Annex B', 2022, tinyurl.com/4m6vux7d.

12 UNSCEAR, 2014, p. 10.

13 M. Rich, 'In a First, Japan Says Fukushima Radiation Caused Worker's Cancer Death', *New York Times*, 6 September 2018, Section A, p. 8, tinyurl.com/mv9tszmf.

14 Those 2,313 deaths were just in Fukushima prefecture. In other prefectures, there were several hundred. See: M. Tsuboi *et al.*, 'Disaster-related deaths after the Fukushima Daiichi nuclear power plant accident', *Environmental Advances*, vol. 8, 100,248, July 2022, tinyurl.com/4rvbxnxv.

15 WHO, '1986–2016: Chernobyl at 30', 25 April 2016, p. 2, tinyurl.com/4wenp35t.

16 The Chernobyl Forum (2005), p. 35.

17 The Chernobyl Forum (2005), p. 35.

18 The Chernobyl Forum (2005), p. 41.

19 The Chernobyl Forum (2005), p. 21.

20 L. Birmingham and D. McNeill (2012), *Strong in the Rain: Surviving Japan's Earthquake, Tsunami, and Fukushima Nuclear Disaster*, Palgrave, p. 87.

21 See, e.g.: S. Ito *et al.*, 'Overview of the pregnancy and birth survey section of the Fukushima Health Management Survey: Focus on mothers' anxieties about radioactive exposure', *Journal of the National Institute of Public Health*, vol. 67, no. 1, pp. 59–70, 2018, tinyurl.com/mrymypa5.

22 Tindale wrote those words in 2015 when he was appointed to the Alvin Weinberg Foundation, which has since ceased to exist. The excerpt is still available on the *Dare to Think* website: 'Stephen Tindale says yes to nuclear', tinyurl.com /2ddt4ms2.

23 G. Monbiot, 'Why Fukushima made me stop worrying and love nuclear power', *Guardian*, 21 March 2011, tinyurl.com /5hxtxtfz.

24 Harding (2018).

25 Yoshihiro Katayama, minister of internal affairs. See: Funabashi, p. 37.

26 TEPCO, 'Fukushima Nuclear Accident Analysis Report', 20 June 2012, p. 94, tinyurl.com/45zz4ncj.

27 See, e.g.: 'Food and Radiation Q&A, Consumer Affairs Agency', p. 12, tinyurl.com/ewu999m5; and Funabashi, p. 87.

28 S. Lotto Persio, '"No One Died From Radiation At Fukushima": IAEA Boss Statement Met With Laughter At COP26', *Forbes*, 4 November 2021, tinyurl.com/bdwmt2pm.

29 Naoto Sekimura, a Professor at the Department of Nuclear Engineering and Management at the University of Tokyo, in: 'Japan warns of nuclear fuel melting after quake damage', *Reuters*, 12 March 2012, tinyurl.com/5fewvcft.

30 M. Raddatz, 'U.S. Officials Alarmed By Japanese Handling of Nuclear Crisis', *ABC News*, 16 March 2011, tinyurl.com/mrxmy84u.

31 D. Jamail, 'Fukushima: It's much worse than you think', *Al Jazeera*, 16 June 2011, tinyurl.com/2fh9b838.

32 'Exodus from Tokyo – 1000s flee poison cloud', *Sun*, 14 March 2011.

33 B. Cox, 'It's scary… but nothing like a nuclear bomb', *Sun*, 15 March 2011. Both examples of the *Sun*'s coverage (collected at tinyurl.com/4vdzrzse) are from: M. McCartney, 'Medicine and the media: Panic about nuclear apocalypse overshadows Japan's real plight', *BMJ*, 22 March 2011, 342:d1845, tinyurl.com/hamhemv9.

34 'Starving Brit Keely: My nightmare trapped in City of Ghosts – Tokyo', *Sun*, 17 March 2011.

35 CBS Evening News, 28 March 1979, tinyurl.com/4fbm77tc.

36 Website, *Bezinningsgroep Energie*, tinyurl.com/56s7nhyu.

37 *NOS Journaal*, 15 March 2011, 12.00, tinyurl.com/2cz6ykmu.

38 S. Ito, 'Suicides stalk Japan disaster zone', *AFP*, 22 August 2011, tinyurl.com/25kza2jx.

39 M. Saito and L. Twaronite, 'Fukushima farmer takes on nuclear plant operator over wife's suicide', *Reuters*, 9 July 2014, tinyurl.com/55hf5kr8.

40 R. Gilhooly, 'Suicides upping casualties from Tohoku catastrophe', *Japan Times*, 23 June 2011, tinyurl.com/y2t5jcce.

41 Ito (2011).

42 Y. Takebayashi *et al.*, 'Characteristics of Disaster-Related Suicide in Fukushima Prefecture After the Nuclear Accident', *Crisis*, vol. 41, no. 6, pp. 475–82, November 2020, tinyurl.com/4vk532hp.

43 'Japan: Nuclear panic is "over-reaction" say scientists', *Channel 4 News*, 17 March 2011, tinyurl.com/yeyjpx2h.

44 G. Knobel, 'Hoe de UU het gezicht werd van een ramp', *DUB*, 20 April 2011, tinyurl.com/3cw46juy.

45 M. Vossen, 'Fukushima: ook een mediaramp', tinyurl.com /2p9dxk8s.

46 *NOS Journaal*, 5 March 2021, 11.00, tinyurl.com/4nxj2fed.

47 E. Timmer, 'Stralingsfobie na ramp Fukushima: "Radioactief gevaar werd overschat"', *De Telegraaf*, 11 March 2021, tinyurl .com/48wmvtsz.

48 ICRP (2009), 'Application of the Commission's Recommendations for the Protection of People in Emergency Exposure Situations', ICRP Publication 109, p. 43, tinyurl.com/yrkknxcp.

49 P. Thomas and J. May, 'Coping after a big nuclear accident', *Process Safety and Environmental Protection*, vol. 112, Part A, November 2017, pp. 1–3, tinyurl.com/yn8hssf5.

50 ICRP (2008), 'Application of the Commission's Recommendations for the Protection of People in Emergency Exposure Situations', p. 57, tinyurl.com/5abjcwb8.

51 IAEA, 'Intervention Criteria in a Nuclear or Radiation Emergency', IAEA, Vienna, 1994, tinyurl.com/yhynmmtu.

52 Harding (2018).

53 A. Gilligan, 'Fukushima: Tokyo was on the brink of nuclear catastrophe, admits former prime minister', *The Telegraph*, 4 March 2016, tinyurl.com/2kb8v486.

Chapter 7

1 Eve (1939), p. 102. Quoted in: R. Rhodes (1986), *The Making of the Atomic Bomb*, Simon & Schuster, p. 44.

2 H.G. Wells (1914), *The Last War: A World Set Free*, p. 59, University of Nebraska Press, tinyurl.com/bdz8t6ft. The book is better known by its original title, *The World Set Free*.

3 C. Nelson (2014), *The Age of Radiance: The Epic Rise and Dramatic Fall of the Atomic Era*, Scribner, p. 80.

4 C. Coleman, 'Atomic super-bomb, made at Oak Ridge, strikes Japan', *Knoxville News Sentinel*, 6 August 1945, p. 1, tinyurl.com/5yfzxyn9.

5 'Rain of Ruin Faced by Japanese As Super Atomic Bomb Loosed', *San Diego Union-Tribune*, 6 August 1945, p. 1, tinyurl.com/5fyz3b9e.

6 Speech, 6 August 1945, tinyurl.com/efutaskt.

7 Declaration, 14 August 1945, tinyurl.com/mufhh9az.

8 Declaration by Pope Pius XII before the Pontifical Academy of Sciences, 8 February 1948, tinyurl.com/v7x68ytz.

9 The Frisch-Peierls Memorandum, written in March 1940, tinyurl.com/3fsyr9d8.

10 The Szilárd Petition to the President of the United States, 17 July 1945, tinyurl.com/2x8xcdma.

11 Henry Stimson in his Memo to President Truman, 25 April 1945, tinyurl.com/bdcwwbt9. Cited in: H. Stimson, 'The Decision to Use the Bomb', *Harper's Magazine*, February 1947.

12 Documentary, *The Decision to Drop the Bomb*, 1965, tinyurl.com/26wwd7r8.

13 'The Bomb That Has Changed the World', *Daily Express*, 7 August 1945, p. 1, tinyurl.com/mme6ze8y.

14 See, e.g., data from Our World in Data: 'War and Peace', tinyurl.com/2p9b84cz.

15 Wells (1914), p. 61.

16 The source remained unnamed. M. Belvedere, 'Trump asks why we can't use nukes', MSNBC, 3 August 2016, tinyurl.com/yctujk6e.

17 Statement, 5 April 2009, in Prague, tinyurl.com/ymped8fk.

18 Current count of nuclear weapons at Statista: tinyurl.com/2p9dxzjv.

19 E. Schlosser, 'The growing dangers of the new nuclear-arms race', *New Yorker*, 24 May 2018, tinyurl.com/33hvt9j2.

20 B. Plackett, 'The Science of Dismantling a Nuclear Bomb', *Inside Science*, 8 February 2019, tinyurl.com/2p8ykya2.

21 G. Cravens (2007), *Power to Save the World: The Truth About Nuclear Energy*, Vintage Books, p. 153.

22 B. von Suttner (1910), *Memoirs of Bertha von Suttner: The Records of an Eventful Life* (Vol. 1 of 2), The Athenaeum Press, p. 210, tinyurl.com/45cvnv5n.

23 See, e.g., K. Joseph, 'Alfred Nobel's Obituary Calling him a "Merchant of Death" Never Happened & Never Inspired the Nobel Prize', *Kathy Loves Physics*, 21 April 2019, tinyurl .com/ut3nrr6w.

24 Von Suttner (1910), p. 437.

25 W. Churchill (1929), *The aftermath: The world crisis: 1918–1928*, Macmillan. Quoted in: J. Mueller (2010), *Atomic Obsession: Nuclear Alarmism from Hiroshima to Al-Qaeda*, Oxford University Press, p. 24.

26 S. Freud (1930), *Civilisation and Its Discontents*, Penguin Books, p. 70. Quoted in: Mueller (2010), p. 24.

27 R. Fredette (1976), *The Sky on Fire: The First Battle of Britain, 1917–1918*, Harcourt Brace Jovanovich, p. 39.

28 Basil Liddell Hart. Cited in: J. Mearsheimer (1988), *Liddell Hart and the Weight of History*, Cornell University Press, p. 90. Quoted in: Mueller (2010), p. 25.

29 J.M. Spaight (1938), *Air Power in the Next War*, G. Bles Publishing, p. 126. Quoted in: G. Quester (1986, original 1966), *Deterrence Before Hiroshima: The Airpower Background of Modern Strategy*, Routledge, p. Xvi.

30 Wells (1914), p. 60.

31 Broadcast TV programme 'Nieuwsuur', 29 April 2016, tinyurl.com/2p82vdzj.

32 See: 'Solid Cancer Risks among Atomic-bomb Survivors', Radiation Effects Research Foundation, tinyurl.com/2p 9fzc3h, and: 'Leukemia Risks among Atomic-bomb

Survivors', Radiation Effects Research Foundation, tinyurl.com/ycxkut7r.

33 'A Jap burns', *Life*, 13 August 1945, p. 34, tinyurl.com /2p96xnn4.

34 Statement, 14 August 1945, tinyurl.com/mufhh9az.

35 A. Cockburn, 'Big Six v. Little Boy', *London Review of Books*, vol. 45, no. 22, 16 November 2023, tinyurl.com/y3n2xz9v.

36 Reception at the Polish Embassy in Moscow, 18 November 1956. See: 'We Will Bury You!', *Time*, 26 November 1956, tinyurl.com/yc5j23eu.

37 Wells (1914), p. 21.

38 Letter to Gertrud Weiss, 6 August 1945. Quoted in: Rhodes (1986), p. 735.

39 Wells (1914), p. 24.

40 Wells (1914), p. 18.

41 Wells (1914), p. 78.

42 Wells (1914), p. 91.

43 Charter of the United Nations, tinyurl.com/2ynw98px.

Chapter 8

1 Lewis Strauss, 16 September 1954, tinyurl.com/ajbafykc.

2 Herbert Kouts, director of reactor safety, AEC, in: 'Hearings Before the Joint Committee on Atomic Energy, Congress of the United States, January 22, 23, 24 and 28, 1974, Part II, volume 1', p. 752, tinyurl.com/4mcrka3f.

3 David Rossin, nuclear research engineer, Commonwealth Edison Company, in: D. Burnham, 'Nuclear Experts Debate "The China Syndrome"', *New York Times*, 18 March 1979, section D, p. 1, tinyurl.com/3a2c69tx.

4 John 'Jack' Herbein, vice president, Metropolitan Edison, press conference, 19 March 1979. A recording can be seen in the documentary *Meltdown at Three Mile Island* (1999), tinyurl.com/yckhymfb.

5 E.F. Schumacher (2011, originally 1973), *Small Is Beautiful: A Study of Economics as if People Mattered*, Vintage Books, p. 115, tinyurl.com/2p9ds6ue.

6 Speech, 7 November 1973, tinyurl.com/4xpnsk4n.

7 Press conference, 17 November 1973, tinyurl.com/bdz59uhx.

8 A 1972 pamphlet by Aktie Strohalm. Seen in: H.J.A. Hofland *et al.* (1983), *Een teken aan de wand: Album van de Nederlandse samenleving 1963–1983*, Bert Bakker, p. 70.

9 The Energy Research and Development Association (ERDA), established in 1975. It merged two years later with the Federal Energy Administration to form the US Department of Energy.

10 G. Cravens (2007), *Power to Save the World: The Truth About Nuclear Energy*, Vintage Books, p. 217.

11 H. Muller, 'Artificial Transmutation of the Gene', *Science*, vol. 66, no. 1,699, pp. 84–7, 22 July 1927, tinyurl.com /2p8w83ua.

12 Nobel Lecture, 12 December 1946, 'The Production of Mutations', tinyurl.com/22aa42wp.

13 Chandra Muller, granddaughter of Hermann Muller, in: T. Green, 'Hermann Muller: A genetics pioneer', *UT News*, 28 September 2007, tinyurl.com/2s6wu5bw.

14 'The Biological Effects of Atomic Radiation: A Report for the Public', 1956, National Academy of Sciences, National Research Council, tinyurl.com/mtzks3. Quotations in the next paragraphs are from pages 2, 3, 7, 20, 23 and 32, plus the Introduction which has no page numbers.

15 See, e.g., E. Calabrese, 'On the origins of the linear no-threshold (LNT) dogma by means of untruths, artful dodges and blind faith', *Environmental Research*, vol. 142, October 2015, pp. 432–42, tinyurl.com/muyzadf4; and R. Adams, 'How did leaders of the Hydrocarbon Establishment build the foundation for radiation fears?', *Atomic Insights*, 21 May 2020, tinyurl.com/sxwvhep5.

16 Warren Weaver.

17 Detlev Bronk.

18 A. Leviero, 'Scientists Term Radiation a Peril to Future of Man', *New York Times*, 13 June 1956, p. 1, tinyurl.com/ yzwy7zns.

19 Arthur Hays Sulzberger.

20 This comes from B. Sacks *et al.*, 'Epidemiology Without
    Biology: False Paradigms, Unfounded Assumptions, and
    Specious Statistics in Radiation Science', in: *Biological Theory*,
    2016, vol. 11, pp. 69–101, tinyurl.com/mrsntrtb.

21 This is based on J. Goldstein and S. Qvist (2019), *A Bright
    Future: How Some Countries Have Solved Climate Change and
    the Rest Can Follow*, Public Affairs, p. 100.

22 UNSCEAR, *Report of the United Nations Scientific Committee
    on the Effects of Atomic Radiation*, 1958, p. 170, tinyurl.com/
    refuvy9h.

23 JCAE, 'The Nature of Radioactive Fallout and Its Effect
    On Man: Summary-analysis of Hearings, May 27–29, and
    June 3–7, 1957', p. 13, August 1957, tinyurl.com/598xp3y7.

24 Quoted in: R.H. Clarke and J. Valentin, 'The History of
    ICRP and the Evolution of its Policies', ICRP Publication
    109, p. 93, 2009, tinyurl.com/mrysu3xm.

25 National   Committee   On   Radiation   Protection   And
    Measurements, 'Maximum Permissible Radiation Exposures
    to Man', *Radiology*, vol. 71, no. 2, August 1958, tinyurl.com
    /2pxs9vsb.

26 UNSCEAR, '2020/2021 Report, Annex B, Levels and
    effects of radiation exposure due to the accident at the
    Fukushima Daiichi Nuclear Power Station: Implications of
    information published since the UNSCEAR 2013 Report',
    2022, p. 5, tinyurl.com/m9rs7ped. UNSCEAR, 'Sources,
    Effects and Risks of Ionizing Radiation, UNSCEAR 2012
    Report to the General Assembly with Scientific Annexes,
    Annex A. Attributing health effects to ionizing radiation
    exposure and inferring risks', 2015, pp. 10, 58, tinyurl.com
    /4u3rntcj.

27 Lauriston   Taylor   in   his   1980   Sievert   lecture: 'Some
    non-scientific influences on radiation protection standards
    and practice', tinyurl.com/23c7apfc.

28 The Japanese professor is Mitsuhei Murata, who advocated
    a ban on nuclear power at the 20th Congressional Briefing

of the Coalition Against Nukes, organised by the Congressional Office of Congressman Kucinich, tinyurl .com/bdhxkths.

29 C. Perrow, 'Five assessments of the Fukushima disaster', *Bulletin of the Atomic Scientists*, 10 March 2014, tinyurl.com /2p9xs93u.

30 C. Perrow, 'Fukushima and the inevitability of accidents', *Bulletin of the Atomic Scientists*, November/December 2011, tinyurl.com/2xch235y.

31 A. Markandya *et al.*, 'Electricity generation and health', *The Lancet*, 370:979–90, 13 September 2007, tinyurl.com/ mutp9b5x.

32 B. Sovacool *et al.*, 'Balancing safety with sustainability: assessing the risk of accidents for modern low-carbon energy systems', *Journal of Cleaner Production*, vol. 112, part 5, pp. 3,952–65, 20 January 2016, tinyurl.com/bdfxz8bc.

33 P. Johnstone *et al.*, 'Nuclear power: Serious risks', *Science*, vol. 354, no. 6,316, p. 1,112, 2 December 2016, tinyurl.com /mry4pcx7.

34 UNSCEAR, 'Sources, Effects and Risks of Ionizing Radiation, UNSCEAR 2013 Report, Volume I, Report to the General Assembly with Scientific Annexes, Scientific Annex A: Levels and effects of radiation exposure due to the nuclear accident after the 2011 great east-Japan earthquake and tsunami', 2014, p. 6, tinyurl.com/mppju5c9.

35 Y. Funabashi (2021), *Meltdown: Inside the Fukushima Nuclear Crisis*, Brookings Institution Press, p. 495.

36 Arthur D. Little Memorial Lecture, 'Physics in the Contemporary World', 25 November 1947.

37 Lecture for Johns Hopkins Applied Physics Laboratory, 'Why is everyone afraid of nuclear power?', 7 February 2020, tinyurl.com/3wznd6wh.

38 E.g. in: E. Calabrese, 'Muller's Nobel Prize lecture: When ideology triumphed over science', *Toxicological Sciences*, vol. 126(1), pp. 1–4, March 2012, tinyurl.com/vmmhydfm; E. Calabrese, 'On the origins of the linear no-threshold (LNT)

dogma by means of untruths, artful dodges and blind faith',
Environmental Research, vol. 142, pp. 432–42, October 2015,
tinyurl.com/489c44az; E. Calabrese, 'Muller's Nobel Prize
research and peer review', *Philosophy, Ethics, and Humanities in
Medicine*, vol. 13(1), p. 15, October 2018, tinyurl.com/2z84n7a9;
and E. Calabrese, 'Muller's Nobel Prize data: Getting the
dose wrong and its significance', *Environmental Research*,
176:108528, September 2019, tinyurl.com/2dy8smre.

39 Letter from Muller to Curt Stern, 14 January 1947. Quoted
in: Calabrese (2015).

40 E. Caspari and C. Stern, 'The influence of chronic irradiation
with gamma-rays at low dosages on the mutation rate in
Drosophila melanogaster', *Genetics*, vol. 33(1), pp. 75–95,
January 1948, tinyurl.com/yeytv76u.

Chapter 9

1 @GretaThunberg, X, 20 August 2021, tinyurl.com/3shyauay.

2 D. Crouch, 'The Swedish 15-year-old who's cutting class to
fight the climate crisis', *Guardian*, 1 September 2018, tinyurl
.com/2d6fdrhk.

3 Speech, UN Climate Action Summit, 23 September 2019,
tinyurl.com/44s9yzd6.

4 T. Bruckner *et al.*, 'Energy Systems', in: 'Climate Change
2014: Mitigation of Climate Change, Contribution of
Working Group III to the Fifth Assessment Report of the
Intergovernmental Panel on Climate Change', Cambridge
University Press, p. 516, tinyurl.com/y96yzc22.

5 UNECE, 'Technology Brief: Nuclear Power', August 2021,
tinyurl.com/mw8rs9xf.

6 IEA, 'Nuclear Power in a Clean Energy System', May 2019,
tinyurl.com/4aadvdk3.

7 S. Arrhenius, 'On the influence of carbonic acid in the air
upon the temperature of the ground', *Philosophical Magazine
and Journal of Science*, vol. 41(251), pp. 237–76, April 1896,
tinyurl.com/yewfafyz.

8   The technical book is *Lehrbuch der kosmischen Physik* (1903, S. Hirzel); the public version is *Världarnas utveckling* (1906, H. Geber), also published as *Worlds in the Making* (1908, Harper).

9   M. Ernman *et al.* (2019), *Our House Is on Fire: Scenes of a Family and a Planet in Crisis*, Allen Lane, p. 146.

10 Emissions are measured in 'carbon dioxide equivalents', also known as: CO_2eq. This means that gases other than CO_2 are weighted by the amount of warming they cause over a 100-year time scale. A good overview of the course of greenhouse gas emissions can be found on Our World in Data's 'Greenhouse gas emissions' page, tinyurl.com/2s43uf3z.

11 S. Schlömer *et al.*, 'Annex III: Technology-specific cost and performance parameters', in: 'Climate Change 2014: Mitigation of Climate Change, Contribution of Working Group iii to the Fifth Assessment Report of the Intergovernmental Panel on Climate Change', 2014, p. 1,335, Cambridge University Press, tinyurl.com/3ezec5bv.

12 Greenpeace Australia website, tinyurl.com/mt7bj6pj.

13 The IEA website features country profiles, showing basic facts about anything from energy consumption to the energy mix.

14 See, e.g., J. Jenkins, 'Historic Paths to Decarbonisation', The Breakthrough Institute, 3 April 2012, tinyurl.com/yc72dbcb; and M. Nelson *et al.*, 'The Power to Decarbonise: Characterising the Impact of Hydroelectricity, Nuclear, Solar, and Wind on the Carbon Intensity of Energy', p. 31, Environmental Progress, November 2017, tinyurl.com/m6peszb9.

15 See, e.g., quote from Jeroen de Haas, top executive of Eneco, in: J.-H. Strop, 'Straks draait het allemaal om de zon en de wind', *De Pers*, 26 March 2012.

16 A.-S. Moreau, 'The sun and the wind will never send us a bill!', *Philonomist*, 25 February 2021, tinyurl.com/4x65u832.

17 Data from Statista: 'Electricity prices for households in Europe in 2021, by country', tinyurl.com/yd7se27n.

18 Margarete Schramböck, speaking as minister for economic affairs, at a meeting of the Competitiveness Council of the Council of the European Union, tinyurl .com/mk9xhmz9.

19 E. Olson, 'A Tale of Two Decarbonizations', The Breakthrough Institute, 1 July 2020, tinyurl.com/kx8yv7j7. Electricity mix in the UK since 1990: tinyurl.com/mt5tx47m. Electricity mix in Germany since 1990: tinyurl.com/ycxwe2z2.

20 As told by Reinhard Urschel, biographer and author of *Gerhard Schröder: Eine Biographie*, on Schröder's 60th birthday, 7 April 2004, in: E. Durak, 'Zum 60. Geburtstag des Bundeskanzlers', *Deutschlandfunk*, tinyurl.com/mr3b5ntf.

21 'In focus: Reducing the EU's dependence on imported fossil fuels', European Commission, 20 April 2022, tinyurl.com/ yhp6md2k.

22 'What's the Lifespan for a Nuclear Reactor? Much Longer Than You Might Think', U.S. Department of Energy, 16 April 2020, tinyurl.com/598p2m3d.

23 IEA, 2019, p. 5.

24 Letter from Sigmar Gabriel to Stefan Lövfen, 13 October 2014, tinyurl.com/yck37ezv.

25 W. van de Velden, 'De energie-expert die de groene droom kan waarmaken', *De Tijd*, 21 November 2020, tinyurl.com/ euthkepx.

26 P. Sertyn, '"De politici moeten niet zo bang zijn voor Electrabel"', *De Standaard*, 12 November 2011, tinyurl.com /ywsabd5.

27 T. Van der Straeten, 'Is de kernuitstap een nieuwe relance of wordt het meer van hetzelfde op de energiemarkt?', *De Tijd*, 5 July 2012, tinyurl.com/3tfrckuf.

28 Blixt website, tinyurl.com/uwanv769.

29 T. Vermeir, 'Hoe moet het verder met onze elektricite-itsproductie nu Doel 3 en Tihange 2 nooit meer zullen heropstarten?', *De Morgen*, 14 June 2014, tinyurl.com/ bdz69wrs.

30 T. Santens, 'Akkoord om alternatieven voor kerncentrales te betalen: bekijk hier wie voor de kosten opdraait', VRT Nieuws, 1 July 2020, tinyurl.com/mphee25j.

31 N. Verhaeghe, 'Subsidies aan gascentrales krijgen groen licht van Europa', VRT Nieuws, 27 August 2021, tinyurl.com/2p83b8e4.

32 @Shell_NatGas, X, 18 January 2018, tinyurl.com/yt6wa7t7.

33 @bp_plc, X, 27 January 2019, tinyurl.com/2fk2282s.

34 Advertising campaign, tinyurl.com/meww4m8y.

35 'Natural Gas, A Partner for Renewable Energy: Report to the G20 Energy and Sustainability Working Group', International Gas Union, May 2015.

36 Groen website, tinyurl.com/4s96arrj.

37 Groen website, tinyurl.com/2thanmet.

38 Tinne Van der Straeten website, tinyurl.com/2p85963p.

39 P. Cobbaert, 'Maak kennis met Tinne Van der Straeten, boegbeeld van Groen en kersvers federaal minister van Energie: "Ik ben zeker: de kerncentrales sluiten in 2025"', De Zondag, 11 October 2020, tinyurl.com/3fpdm4wp.

40 R. Liekens, '"Ik garandeer iedereen: het licht zal niet uitvallen"', Humo, 7 December 2020, tinyurl.com/p45utf4k.

41 '5 vragen aan de minister van Energie', Izen, tinyurl.com/25rrt4r2.

42 Interview by Terzake, VRT, 17 November 2020, tinyurl.com/d32sy4fz.

43 C. Gürsan et al., 'The systematic impact of a transition fuel: Does natural gas help or hinder the energy transition?', Renewable and Sustainable Energy Reviews, vol. 138, March 2021, tinyurl.com/tpzwc7tr.

44 C. Moore, 'Vision or division? What do National Energy and Climate Plans tell us about the EU power sector in 2030?', Ember, November 2020, tinyurl.com/2p8e26fk.

45 In book form, the report appeared as: C. Wilson et al. (1971), Inadvertent Climate Modification: Report of Conference, Study of Man's Impact on Climate (SMIC), MIT Press.

46 National Research Council (1977), *Energy and Climate: Studies in Geophysics*, The National Academies Press.

47 W. Sullivan, 'Scientists Fear Heavy Use of Coal May Bring Adverse Shift in Climate', *New York Times*, 25 July 1977, p. 1, tinyurl.com/4ena93v5.

48 'CO_2 Pollution May Change the Fuel Mix', *Business Week*, 8 August 1977, p. 25.

49 Television speech, 'The Crisis of Confidence', 15 July 1979, tinyurl.com/44fvdh56.

50 K. Teltsch, 'Carter Proposes a Nuclear Limit', *New York Times*, 14 May 1976, p. 1, tinyurl.com/ycydf53j. The article lists seven figures with whom the speech was discussed: Zbigniew Brzezinski, W. Averell Harriman, Paul Nitze, Gerard Smith, Cyrus Vance, Paul Warnke and Charles Yost.

51 Data from Energy Information Administration, tinyurl.com /ecf3p2m5. By 2021, it would amount to 2.8 per cent. During those years, the share of fossil fuels also went up.

52 W. Sullivan, 'Study Finds Warming Trend That Could Raise Sea Levels', *New York Times*, 22 August 1981, tinyurl .com/467mhu2t.

53 Statement to Congress, 23 June 1988, tinyurl.com/2mddpjjh.

54 P. Shabecoff, 'Global Warming Has Begun, Expert Tells Senate', *New York Times*, 24 June 1988, p. 1, tinyurl.com /2u5faf6s.

55 Shabecoff (1988).

56 J. Noble Wilford, 'Broeikaseffect: er is al geen weg meer terug', *de Volkskrant*, 3 September 1988, p. 33, tinyurl.com /234kndkn. This was a translation of an article that ran in the 23 August edition of the *New York Times*, 'His Bold Statement Transforms the Debate On Greenhouse Effect', tinyurl.com/5ypj7bx4.

57 Speech, 8 November 1989, tinyurl.com/2hvdn7yc. Video recording: tinyurl.com/2vxzep8s.

58 @janhaverkamp, X, 15 February 2019, tinyurl.com/mreutpbp.

59 @digitalwerber, X, 24 September 2021, tinyurl.com/ mhch8r78.

60 D. van Dam, 'Pleidooi voor kernenergie bij Klimaatmars valt verkeerd: "Pakken lekgeprikt"', *De Telegraaf*, 6 November 2021, tinyurl.com/mpjuswxm. Personal note: I was there that day, joining the pro-nuclear group.

61 E.g., in: M. Jacobson *et al.*, 'Low-cost Solution to the Grid Reliability Problem With 100% Penetration of Intermittent Wind, Water, and Solar For All Purposes', *Proceedings of the National Academy of Sciences of the United States*, vol. 112, no. 49, 8 December 2015, tinyurl.com/mj744tp.

62 J. Hansen *et al.*, 'Nuclear power paves the only viable path forward on climate change', *Guardian*, 3 December 2015, tinyurl.com/yzx763hm.

63 N. Oreskes, 'There is a new form of climate denialism to look out for – so don't celebrate yet', *Guardian*, 16 December 2015, tinyurl.com/5hbrk98p.

64 Hansen *et al.* (2015).

65 Ernman (2019), p. 259.

66 C. van Mersbergen, 'Brussel houdt nieuwe kerncentrales niet tegen', *Algemeen Dagblad*, 26 October 2020, tinyurl.com /muk7jkuz.

67 Nelson is founder and managing director of Radiant Energy, and a frequent guest at the Decouple podcast. Personal correspondence, 27 October 2023.

68 J. Hoenders and B. Hettema, 'John en Wim vochten 50 jaar geleden al tegen kernenergie, nieuwe kerncentrales vinden ze nog steeds onaanvaardbaar', *EenVandaag*, 4 January 2022, tinyurl.com/2vwekjbb.

69 G. Monbiot, 'The Moral Case for Nuclear Power', monbiot .com, 8 August 2011, tinyurl.com/mr269wez.

70 Speech, 8 November 1989.

71 Presentation by Alvin Weinberg, 'Nuclear energy and the greenhouse effect', at the Midwest Energy Consortium Symposium, Chicago, 6 November 1989. Excerpted in: A. Weinberg (1992), *Nuclear Reactions: Science and Trans-Science*, p. 318, The American Institute of Physics.

72 'Environmental Implications of the New Energy Plan: Hearing Before the Subcommittee on the Environment and the Atmosphere of the Committee on Science and Technology, U.S. House of Representatives, Ninety-fifth Congress, First Session, June 8, 9; July 21, 26, 27; and September 27, 28, 29, 1977', p. 456, tinyurl.com/5ebydabk.

73 A. Weinberg, 'Global Effects of Man's Production of Energy', *Science*, vol. 186, no. 4,160, p. 205, 18 October 1974, tinyurl.com/hrrrjde6.

74 C. Whittle *et al.* (1976), 'Economic and Environmental Implications of a US Nuclear Moratorium, 1985–2010', Institute for Energy Analysis, tinyurl.com/2nmc76fz.

75 D. Meadows (1972), *The Limits to Growth: A Report for the Club of Rome's Project on the Predicament of Mankind*, Potomac Associate Books, p. 73, tinyurl.com/2482nt57.

76 A. Weinberg (1994), *The First Nuclear Era: The Life and Times of a Technological Fixer*, Springer, pp. 236–7.

77 @rtgniessen, X, 6 November 2021, tinyurl.com/vccbjzke.

Chapter 10

1 A. Kaufman, 'Finland Is On The Brink Of A Nuclear Power "Game Changer"', *HuffPost*, 23 April 2022, tinyurl.com /4bvryn76.

2 Greenpeace video recording: tinyurl.com/3r9zrv6x.

3 C. Hebel *et al.*, 'Report to the American Physical Society by the study group on nuclear fuel cycles and waste management', *Reviews of Modern Physics*, vol. 50, no. 1, vol. II, January 1978, pp. S5–6, tinyurl.com/t3r9ycew.

4 M. Weber, 'Summary and Observations of the Conference for a Nuclear Free 1990s, 26–27 April 1991', 30 April 1991, tinyurl.com/bdhvxfps.

5 'Wat zou jij doen met vijf zwembaden aan kernafval? Verzet je nu tegen geologische berging in ons land', Groen website, 19 May 2020, tinyurl.com/388j8au3.

6 Press release, Greenpeace, 'Greenpeace: Kernafval nog 24.000 jaar gevaarlijk', 29 September 2004, tinyurl.com/4hkv24d5.

7 S. Brand (2010), *Whole Earth Discipline: Why Dense Cities, Nuclear Power, Genetically Modified Crops, Restored Wildlands, Radical Science and Geoengineering Are Essential*, p. 78, Atlantis.

8 Brand (2010), p. 78.

9 Or a golf ball. See: B. Brook, 'Golf balls and elephants: Energy density in 9 seconds', *Brave New Climate*, 22 June 2011, tinyurl.com/ycxakpwn.

10 Or a brick. See: 'What is nuclear waste and what do we do with it?', World Nuclear Association, tinyurl.com/327y5jpx.

11 It is 29,000 cubic metres (over a million cubic feet) worldwide, according to the IAEA in a 2022 report, 'Status and Trends in Spent Fuel and Radioactive Waste Management', p. 50, tinyurl.com/yckw57t3. A football field is 105 x 68m (340 x 220ft).

12 Rough estimate for nuclear waste based on 25 to 30 tonnes of highly radioactive waste from a typical large 1 GW nuclear reactor, at 440 nuclear reactors around the world. Source: 'Radioactive Waste: Myths and Realities', World Nuclear Association, January 2022, tinyurl.com/4mhztws2. E-waste: 50 million tonnes a year, according to the Platform for Accelerating the Circular Economy, which includes the UN Environment Programme, 'A New Circular Vision for Electronics: Time for a Global Reboot', January 2019, World Economic Forum, tinyurl.com/38f73pzk. Household waste: at least 2 billion tonnes per year, according to S. Kaza *et al.* (2018), *What a Waste 2.0: A Global Snapshot of Solid Waste Management to 2050*, The World Bank, tinyurl .com/mr3pkesuk.

13 Documentary by Decouple Media, 'How I Learned to Stop Worrying and Love Nuclear Waste' (2023), available on YouTube, tinyurl.com/yu3wyaw8.

14 DOE Openness (1988), *Human Radiation Experiments: ACHRE Report*, Chapter 5: 'The Manhattan District Experiments', tinyurl.com/bdzmhskn.

15 See, e.g., the heading 'Toxicity' at Wikipedia's page on plutonium: tinyurl.com/24upnss2.

16 He wrote about this at several places in, e.g.: B. Cohen (1990), *The Nuclear Energy Option: An Alternative for the 90s*, Springer, tinyurl.com/2zkvdfzx.

17 J. Devanney (2020), *Why Nuclear Power Has Been a Flop*, BookBaby, p. 178.

18 R. Partanen and J. Korhonen (2020), *The Dark Horse: Nuclear Power and Climate Change*, Let's Think Again Media, p. 134.

19 M. Sintubin, 'Durf kernafval toevertrouwen aan Planeet Aarde', *Ecomodernisme.be*, 2 October 2020, tinyurl.com/yke7s4an.

20 Press release, Bundesgesellschaft für Endlagerung, '90 Teilgebiete in ganz Deutschland', 28 September 2020, tinyurl.com/bdfzznrx.

21 Press release, De Nationale Geologiske Undersøgelser for Danmark og Grønland, 'Danmarks undergrund evalueret til deponering af radioaktivt affald', 19 January 2022, tinyurl.com/2p957dkn.

22 See, e.g., T. Hjerpe *et al.* (2010), *Biosphere Assessment Report 2009*, Posiva, tinyurl.com/22j6bwh2, and J.M. Korhonen, '"Graph" of the week: What happens if nuclear waste repository leaks?', The unpublished notebooks of J.M. Korhonen, 15 August 2013, tinyurl.com/mw7cnad7.

23 A. Lempinen and M. Silvan-Lempinen (2011), 'Reverse Logic: Safety of Spent Nuclear Fuel Disposal', Greenpeace International, p. 38. Quoted in: R. Partanen and J. Korhonen (2015), *Climate Gamble: Is Anti-Nuclear Activism Endangering Our Future?*, p. 66.

24 A. Gentleman, 'Fatality fuels anti-nuclear protest', *Guardian*, 10 November 2004, tinyurl.com/mw7eet7.

25 S. Johnson and J. Ahlander, 'Sweden approves plan to bury spent nuclear fuel for 100,000 years', *Nasdaq*, 27 January 2022, tinyurl.com/yn4vhnv9.

26 S. Tuler and T. Webler, 'A better way to store nuclear waste: Ask for consent', *Bulletin of the Atomic Scientists*, 21 April 2021, tinyurl.com/yc7y3vxj.

27 H. Fountain, 'On Nuclear Waste, Finland Shows U.S. How It Can Be Done', *New York Times*, 9 June 2017, tinyurl.com/yckesr7p.

28 A. Jaspers, 'Kernenergie is wél een goed alternatief voor fossiel, nu het probleem van radioactief afval is opgelost', *De Correspondent*, 8 November 2019, tinyurl.com/294mvtam.

29 Press release, IAEA, 'Finland's Spent Fuel Repository a "Game Changer" for the Nuclear Industry, Director General Grossi Says', 26 November 2020, tinyurl.com/f4hreczd.

30 M. Sintubin, 'Kernafval, het non-argument in de discussie rond de kernuitstap', *De Morgen*, 15 December 2017, tinyurl.com/y6d8j74j.

31 Quotation from James Lovelock, in Brand (2010), p. 106.

32 M. Verplancke, 'Stop dat sentimentele gejank over hongerige ijsberen', *De Morgen*, 27 March 2018, tinyurl.com/mr3hxrp8.

33 T. Morton (2010), *The Ecological Thought*, Harvard University Press, p. 132, tinyurl.com/29r36rsh.

34 B. Latour, 'Love Your Monsters', *Breakthrough Journal*, no. 2, autumn 2011, p. 21, tinyurl.com/2p8spsrd.

35 M. Lynas *et al.*, 'What a waste: How fast-fission power can provide clean energy from nuclear waste', RePlanet, 2023, tinyurl.com/yx994wsk.

36 Report of the World Commission on Environment and Development (1987), *Our Common Future*, Oxford University Press, tinyurl.com/4wbfzsjv.

37 Websites of the European Commission: 'Circular economy action plan', tinyurl.com/54ke47se, and 'The European Green Deal', tinyurl.com/bdmnskep.

Chapter 11

1 M. Visscher, 'Gaat thorium de wereld redden?', *Vrij Nederland*, 19 December 2017, tinyurl.com/2p8b396r.

2 Wyoming PBS, tinyurl.com/2dct8kbk.

3 Decouple podcast, 'Michael Shellenberger: A Heretic Among Heretics', 11 April 2021, tinyurl.com/3xbrrrtt.

4 'Hearings Before Subcommittees of the Joint Committee on Atomic Energy, Congress of the United States, Eighty-fifth Congress, First Session on Progress Report on Naval Reactor Program and Shippingport Project, March 7 and April 12, 1957', p. 70, tinyurl.com/bdzxa7kc.

5 Data from Our World In Data: 'Access to Energy', tinyurl.com/48k67ecj.

6 Personal interview, 8 June 2022, supplemented by: P. Mthombeni, 'Nuclear energy is critical to Africa's agenda for sustainable development', *World Nuclear News*, 6 January 2022, tinyurl.com/ycx555j4; and P. Mthombeni, 'Nuclear technology for Africa's Agenda for Sustainable Development', *Executive Intelligence Review*, 22 April 2022, tinyurl.com/5a4p47n4.

7 Africa Progress Panel, 'Power, People, Planet: Seizing Africa's Energy and Climate Opportunities', Africa Progress Report 2015, tinyurl.com/bdf5n7sv.

8 See, e.g., IEA, 'World Energy Outlook 2021', tinyurl.com/89uzvuyr; and EIA, 'International Energy Outlook 2021', tinyurl.com/2a3x85yb.

9 'The future of cooling: Opportunities for energy-efficient air conditioning', IEA, 2018, tinyurl.com/r7h54u6x.

10 World Bank website, tinyurl.com/mv9yxxn9, contrasted with: '10 Things You Didn't Know About the World Bank Group's Work on Climate Change', The World Bank, 29 October 2021, tinyurl.com/mwh89yw3.

11 'G7 Climate, Energy and Environment Ministers' Communiqué', 27 May 2022, tinyurl.com/3b28h5td.

12 'Statement on International Public Support for the Clean Energy Transition', 4 November 2021, tinyurl.com/4abjumsd.

13 F. Harvey and P. Greenfield, 'Twenty countries pledge end to finance for overseas fossil fuel projects', *Guardian*, 3 November 2021, tinyurl.com/mrxncwzw.

14 V. Ramachandran, 'Rich Countries' Climate Policies Are Colonialism in Green', *Foreign Policy*, 3 November 2021, tinyurl.com/4yckpd66.

15 R. Milne, 'Drilling shutdown would mean end of green transition, Norway PM warns', *Financial Times*, 25 October 2021, tinyurl.com/4ewyjzus.

16 S. Saran, 'Enough preaching about climate, it's time for "just" action', Observer Research Foundation, 31 March 2021, tinyurl.com/mwj6ppv9.

17 Data from Our World In Data: 'Cumulative CO_2 emissions', tinyurl.com/2setbb3n.

18 Most up-to-date status of emerging nuclear energy countries at World Nuclear Association, tinyurl.com/y8r2fhd8.

19 UNECE, 'Application of the United Nations Framework Classification for Resources and the United Nations Resource Management System: Use of Nuclear Fuel Resources for Sustainable Development – Entry Pathways', 2021, p. 6, tinyurl.com/4h4ardph.

20 IEA, 'World Energy Outlook 2021', p. 297, tinyurl.com /4akd7uwe. Most up-to-date status of plans for reactors world-wide at World Nuclear Association, tinyurl.com/2y3vau5s.

21 R. Leppert and B. Kennedy, 'Growing share of Americans favor more nuclear power', Pew Research Center website, 18 August 2023, tinyurl.com/3apcbh8j.

22 'Meer Nederlanders voorstander van kernenergie', Centraal Bureau voor de Statistiek, 28 November 2023, tinyurl.com /5anjfxz6

23 'Surge in Public Support for Nuclear Power in Poland', Centrum Badania Opinii Społecznej, CBONews, Newsletter 38, 2022, tinyurl.com/39ubdx33.

24 'Opiniepeiling: Belgen overtuigd voorstander van bestaande én nieuwe kernenergie', Nucleair Forum, 6 November 2023, tinyurl.com/2xtvfu68.

25 'More than half want to build new reactors if needed', Analysgruppen website, 10 April 2022, tinyurl.com/ mrxmpfps.

26 S. Johnson, 'Sweden plans new nuclear reactors by 2035, will share costs', *Reuters*, 16 November 2023, tinyurl.com/ mrwzdv2t.

27 J. Visser, 'Zweden wil komende twintig jaar fors meer kerncentrales bouwen', *de Volkskrant*, 16 November 2023, tinyurl.com/477t8778.

28 J. Lovelock, 'Nuclear power is the only green solution', *Independent*, 24 May 2004, tinyurl.com/3xz5rhem.

29 G. Monbiot, 'The nuclear winter draws near', *Guardian*, 30 March 2000, tinyurl.com/5n6ubsys.

30 G. Monbiot, 'Why Fukushima made me stop worrying and love nuclear power', *Guardian*, 21 March 2011, tinyurl.com /5hxtxtfz.

31 G. Monbiot, 'The farce of the Hinkley C nuclear reactor will haunt Britain for decades', *Guardian*, 22 October 2013, tinyurl.com/mr38wtxn.

32 Monbiot wrote this with two other pro-nuclear environmentalists, Chris Goodall and Mark Lynas: 'We are pro-nuclear, but Hinkley C must be scrapped', *Guardian*, 18 September 2015, tinyurl.com/dmue7zzy.

33 Visscher, 2017.

34 G. Knobel, 'Hoe de UU het gezicht werd van een ramp', *DUB*, 20 April 2011, tinyurl.com/3cw46juy.

35 Greta Thunberg, Facebook, 17 March 2019, tinyurl.com/ ynke7jxc.

36 Speech, Youth4Climate, Milan, 28 September 2021, tinyurl .com/46zf227m.

37 M. Ernman *et al.* (2019), *Our House Is on Fire: Scenes of a Family and a Planet in Crisis*, Allen Lane, p. 87.

38 Speech, R20 Austrian World Summit, Vienna, 28 May 2019, tinyurl.com/3wpsr7pz.

39 Speech, United Nations Climate Change Conference Summit, Katowice, Poland, 15 December 2018, tinyurl.com/ mppm6ewh.

40 N. Klein (2014), *This Changes Everything: Capitalism vs the Climate*, Alfred A. Knopf Canada, p. 16.

41 This Changes Everything website, tinyurl.com/5n7rk8tv.

42 Klein (2014), p. 38.

43 Klein (2014), p. 136.

44 Francis, *Laudato Si'*, verse 161, 24 May 2015, tinyurl.com/ p7c9dap5.

45 Vatican, 'Naomi Klein: "To change everything, we need everyone"', *La Stampa*, 1 July 2015, tinyurl.com/2p9ftfud.

46 Rome Reports, 17 April 2019, tinyurl.com/9a6ebpxw.

47 'Renewable Energy Mythbuster #2', Greenpeace International, Facebook, 27 December 2018, tinyurl.com/ nhdh3f4p.

48 Ernman (2019), p. 222.

49 Speech, UN Climate Action Summit, New York, 23 September 2019, tinyurl.com/44s9yzd6.

50 Speech (23 September 2019).

51 E.F. Schumacher (2011, originally 1973), *Small Is Beautiful: A Study of Economics as if People Mattered*, Vintage Books, p. 46, tinyurl.com/2p9ds6ue.

52 Schumacher (1973), p. 147.

53 C.P. Snow, Rede Lecture, 7 May 1959, p. 12, tinyurl.com /2xwvpcay.

54 Snow (1959), p. 13.

55 E. Zola (1901), *Travail*, tinyurl.com/4fkb4j9p.

56 P. Morand (1931), *1900 A.D.*, William Farquhar Payson, p. 65.

57 Facebook (2019).

58 Ernman (2019), p. 181.

59 Ernman (2019), p. 260.

60 Klein (2014), p. 341.

61 Francis (2015), verse 66.

62 Francis (2015), verse 23.

63 Francis (2015), verse 202.

64 Articles of association, Stichting Greenpeace Council, as approved at the 2016 Annual General Meeting, tinyurl.com /yx3tpp24.

65 Deed, Amendment of the articles of association, Stichting Greenpeace Nederland, 12 October 2018, tinyurl.com /36hej6hh.

66 Speech, UN Climate Action Summit, New York, 23 September 2019. Entire recording: tinyurl.com/2s9drecu.

67 Klein (2014), p. 249.
68 Interview, Covering Climate Now, 12 October 2021, tinyurl.com/5n78hk6w.
69 Vatican (2015).
70 Speech, acceptance of Alumni Award 1966, University of Chicago, tinyurl.com/47tvkrdd.
71 T. Trenn, 'The Central Role of Energy in Soddy's Holistic and Critical Approach to Nuclear Science, Economics and Social Responsibility', in: G. Kaufman (ed.) (1986), *Frederick Soddy (1877–1956): Early Pioneer in Radiochemistry*, D. Reidel Publishing, p. 185. Quoted in: Jenkin (2011).
72 'Atomic Education Urged by Einstein', *New York Times*, 25 May 1946, p. 11, tinyurl.com/y9wexmx5.

Epilogue

1 M. Visscher, 'Verontrustende terugkeer kernenergie', *Ode*, september/oktober 2000, tinyurl.com/5bhnxe42.
2 M. Visscher (2018), *De energietransitie: Naar een fossielvrije toekomst, maar hoe?*, Nieuw Amsterdam, p. 115.
3 A. Blair, 'Russian soldier "dies from radiation poisoning" after Vlad's troops dug trenches at Chernobyl before fleeing to Belarus', *Sun*, 1 April 2022, tinyurl.com/yc6ktcbz.
4 See e.g., M.D. Wood *et al.*, 'Chornobyl radiation spikes are not due to military vehicles disturbing soil', *Journal of Environmental Radioactivity*, vol. 265, September 2023, 107220, tinyurl.com/2p3ymabc, and K. Zetter, 'The Mystery of Chernobyl's Post-Invasion Radiation Spikes', *Wired*, 7 August 2023, tinyurl.com/y3jtzvm2.
5 A. Koshiw and J. Rankin, 'Attack on Ukraine nuclear plant "suicidal", says UN chief as he urges access to site', *Guardian*, 8 August 2022, tinyurl.com/3ekehm27.
6 'Putin allows UN inspectors to visit Zaporizhzhia nuclear plant', Al Jazeera, 19 August 2022, tinyurl.com/ymwzd6za.
7 Press release, IAEA, 'Director General Grossi Alarmed by Shelling at Ukraine NPP, says IAEA Mission Vital for

Nuclear Safety and Security', 6 August 2022, tinyurl.com /2p8v9t9j.

8 D. Castelvecchi, 'Ukraine nuclear power plant attack: scientists assess the risks', *Nature*, 4 March 2022, tinyurl.com /myxt27wf.

9 M. Sangiorgi *et al.* (2016), 'In-Vessel Melt Retention (IVMR) Analysis of a VVER-1000 NNP', JRC Technical Reports, European Union, tinyurl.com/4832jhxz.

10 IAEA, 'Nuclear Safety, Security and Safeguards in Ukraine, 2nd Summary Report by the Director General, 28 April–5 September 2022', tinyurl.com/4nmtjz9t.

11 Briefing, UN Security Council, 6 September 2022, tinyurl .com/3knpfd2c.

12 'Atoomagentschap: Veiligheidszone rond kerncentrale Zaporizja nodig, nucleaire catastrofe dreigt', *Algemeen Dagblad*, 6 September 2022, tinyurl.com/26v7xh9b.

13 'Atoomwaakhond: "Kernramp dreigt bij Oekraïense centrale Zaporizja"', *De Telegraaf*, 6 September 2022, tinyurl .com/a7cstv6t.

Index